Textile

EDITED BY
CATHERINE HARPER
AND DORAN ROSS

THE JOURNAL OF
CLOTH AND CULTURE

VOLUME 9
ISSUE 3
NOVEMBER 2011

ORDERING INFORMATION

Three Issues per volume. One volume per annum.
2011: Volume 9
ONLINE
www.bergpublishers.com
BY MAIL
Berg Publishers
C/O Turpin Distribution Services
Pegasus Drive
Stratton Business Park
Biggleswade
Bedfordshire SG18 8TQ
UK
BY FAX
+ 44 (0)1767 601640
BY TELEPHONE
+ 44 (0)1767 604951
For Subscription Enquiries
email custserv@turpin-distribution.com
ENQUIRIES
Editorial: Julia Hall
email julia.hall@bloomsbury.com
Production: Ken Bruce
email ken.bruce@bloomsbury.com
Advertising: Ellie Graves
email eleanor.graves@bloomsbury.com
SUBSCRIPTION RATES
Institutional
Print and online: 1 year: £170/US$331; 2 year: £272/US$530
Online only: 1 year: £144/US$282; 2 year: £231/US$451
Individual
Print: 1 year: £49/US$85; 2 year: £78/US$136
Full color images available online
Access your electronic subscription through
www.ingentaconnect.com
Berg Publishers is an imprint of
Bloomsbury Publishing Plc.

AIMS AND SCOPE

Cloth accesses an astonishingly broad range of human experiences. The raw material from which things are made, it has various associations: sensual, somatic, decorative, functional and ritual. Yet although textiles are part of our everyday lives, their very familiarity and accessibility belie a complex set of histories, and invite a range of speculations about their personal, social and cultural meanings. This ability to move within and reference multiple sites gives textiles their potency.

This journal brings together research in textiles in an innovative and distinctive academic forum for all those who share a multifaceted view of textiles within an expanded feld. Representing a dynamic and wide-ranging set of critical practices, it provides a platform for points of departure between art and craft; gender and identity; cloth, body and architecture; labor and technology; techno-design and practice—all situated within the broader contexts of material and visual culture.

Textile invites submissions informed by technology and visual media, history and cultural theory; anthropology; philosophy; political economy and psychoanalysis. It draws on a range of artistic practices, studio and digital work, manufacture and object production.

Berg Publishers is a member of CrossRef

SUBMISSIONS

Should you have a topic you would like us to consider, please send an abstract of 300–500 words to one of the editors. Notes for Contributors can be found at the back of the journal and style guidelines are available by emailing ken.bruce@bloomsbury.com or from the Berg website (www.bergpublishers.com).

ISSN: 1475-9756

www.bergpublishers.com

Textile is indexed by Abstracts in Anthropology; AIO (Anthropological Index Online); ART Bibliographies Modern; British Humanities Index; Current Contents/Arts and Humanities; DAAI (Design and Applied Arts Index); IBR (International Bibliography of Book Reviews of Scholarly Literature in the Humanities and Social Sciences); IBSS (International Bibliography of the Social Sciences); IBZ (International Bibliography of Periodical Literature on the Humanities and Social Sciences); ISI Arts and Humanities Citation Index; Scopus; World Textiles.

Contents

EDITORS

Catherine Harper
School of Architecture and Design
University of Brighton
Grand Parade
Brighton BN2 4AY
UK
Catherine.Harper@brighton.ac.uk

Doran Ross
Fowler Museum at UCLA
308 Charles Young Drive
Los Angeles, CA 90095-1549
USA
dross@arts.ucla.edu

CALL FOR PROPOSALS

Textiles that Changed the World Series

Textiles have had a profound impact on the world in a multitude of ways—from the global economy to the practical and aesthetic properties that subtly shape our everyday lives. This exciting series chronicles the cultural life of individual textiles through sustained, book-length examinations. Pioneering in approach, the series focuses on historical, social and cultural issues and the myriad ways in which textiles ramify meaning. Each book is devoted to an individual textile, fiber or dye that characterizes a particular type of cloth. Books are handsomely illustrated with color as well as black-and-white photographs.

Titles published and forthcoming in this series

Jonathan Faiers, *Tartan* (2008)
Willow G. Mullins, *Felt* (2009)
Beverly Lemire, *Cotton* (2011)
Fiona Anderson, *Tweed*

Proposals are invited for additions to this series. Single-authored books rather than edited works are preferred. Please contact the Series Editor for further information or submit to her the following:

- A short (500-word) summary of the proposed book
- A table of contents and detailed chapter summaries
- An overview of any competing or complementary books
- Biographical details/short CV.

CONTACT:
Series Editor: Linda Welters, Department of Textiles, Fashion Merchandising, and Design, University of Rhode Island; lwelters@uri.edu

TEXTILES
Critical and Primary Sources

Edited by Catherine Harper

The first multi-volume reference work to anthologize writing on textiles in the humanities and social sciences.

Textiles: Critical and Primary Sources is a major multi-volume reference work that draws together 80 seminal texts on textiles.

Textile culture stretches geographic, historical, methodological and disciplinary boundaries, and defies chronological ordering. The contents are therefore gathered into four thematic collections dealing with history and curation; production and sustainability; science and technology; and identity, each supported by an introductory editorial essay that serves to critique and supplement each textual collection and theme.

Textiles: Critical and Primary Sources is a key scholarly resource for any researchers involved in the study of textiles, as well as associated subjects, including studies in dress, costume and fashion; feminism and gender; art, design and cultural history; and sociology and anthropology.

Catherine Harper is Dean of the School of Arts and Digital Industries at the University of East London.

Volume 1: History and Curation

Volume 2: Production (including Sustainability)

Volume 3: Science and Technology

Volume 4: Identity

INTRODUCTORY OFFER: £495.00 until 30 April 2012*

February 2012
1,600pp • 244 x 172 mm
HB Set: 978 0 85785 035 5   £495.00
(*price from 01 May 2012: £550.00)

www.bergpublishers.com

New Rules for Old Gems: Can El Salvador Sustain and Develop Home Grown Design?

Abstract

This article evaluates the state of the current textile industry in El Salvador. Its garment industry greatly depends on the maquiladoras, which refers to the practice of millers charging a maquila, or "miller's portion" for processing other people's grain. However, with the threat of booming economies such as India and China offering even lower labor costs and faster production times in factories ten times bigger, the garment industry is a bleak one these days. Around 75 percent of the original factories have already closed down, leaving around 30,000 unemployed Salvadorians. The same is true for other countries in Latin America, especially Mexico and the rest of Central America. The possibilities for the industry to survive and thrive are juxtaposed with successful craft- and handmade-driven businesses in Brazil and the USA. The analysis uncovers other underlying issues such as bootlegging and monetary consignments from abroad as a deterrent for the development of Salvadorian design. The article presents a strategy that not only combines the mass-produced with handmade elements but also empowers women in the Salvadorian economy.

Keywords: maquiladora, Latin America, garment industry, handmade, women

CAROLINA GOMEZ-AUBERT

Born in El Salvador, Carolina Gomez-Aubert is a textile designer and social entrepreneur. She studied Textile Design at Chelsea College of Art and Design, specializing in constructed textiles and surfaces. She lives and works in London and continues to develop "Lunamano," a cooperative of Salvadorian women working with material rescue and redesign.

Textile, Volume 9, Issue 3, pp. 288–307
DOI: 10.2752/175183511X13173703491072
Reprints available directly from the Publishers.
Photocopying permitted by licence only.
© 2011 Berg. Printed in the United Kingdom.

New Rules for Old Gems: Can El Salvador Sustain and Develop Home Grown Design?

Introduction

Thick clusters of tags piled up anywhere the International Monetary Fund or other capitalist colonizing forces have travelled. (SubRosa 2007: 82)

In May 2005, SubRosa, a group of six women artists, exhibited at *The Interventionists: Art in the Social Sphere*, in the Massachusetts Museum of Contemporary Art. Their piece consisted of visitors trimming off the tags from their own clothing, pinning them to a map at the geographical point of manufacture or assembly. It was the "First World" nations, like the USA, Canada, and Western European countries, that remained tagless. Ironic, considering it is these nations that sell the products manufactured and assembled by the "Third World" nations.

The place where it all happens in El Salvador is a maquiladora. The word "conjures up images of massive warehouses, women working long hours for wages that will barely cover the basics and unsafe working conditions" (Howard 2007: 31). In theory though, a maquiladora is much more. It is a factory that imports materials and equipment on a tariff basis for assembly or manufacturing and then re-exports the assembled product, usually to the originating country (*Encyclopaedia Britannica Online* 2007a).

The economic logic behind relocating businesses is the reduction of costs. If some people can sell their skills cheaper than others, those people have the comparative advantage. According to David Ricardo (2006) in his 1817 book *On the Principles of Political Economy and Taxation* this would benefit both countries by increasing their Gross Domestic Product, the total market value of goods and services produced by the country's economy in a given period (*Encyclopaedia Britannica Online* 2007c). For the USA, the owner of 90 percent of the maquiladoras of Latin America this is true, but for the countries that house these factories it is not. Why?

Since the recent boom of economies such as China and India, countries like El Salvador have seen an immense shift in the amount of money invested within. Now the USA chooses much cheaper labor costs, with faster production results in factories almost ten times bigger than those in El Salvador.

So what happens to these now unemployed semi-skilled Salvadorians? What happens to the abandoned infrastructure and machinery? To what extent are countries, like the USA (the employer), responsible for the Salvadorian people (the employee)?

Maybe the solution lies within the Poko Pano case study. Poko Pano is a Brazilian swimwear company that employs people to embellish with handiwork the pieces they produce, assemble, and sell in Sao Paolo and then export. Coopa Roca and the former Project Alabama, now known as Alabama Chanin, are other examples of this kind of local craft applied to the clothing industry. However, Project Alabama was not as successful. They ended up moving their production to India, as they did not have enough labor and capital to cope with production demands.

Is it time for the smallest country in Central America to begin a vertical production scheme, which will provide decent salaries, respectable working conditions, and adequate training for its people?

I intend to answer these questions not only by evaluating the successes and failures of the aforementioned case studies, but also through interviews with Ricardo Escobar, a former owner of a maquiladora in San Salvador and current member of ASIC (Salvadorian Association of Dressmakers Industry); a conversation with the general manager, Samuel Cerna, of "Los Capellanos," a maquiladora in San Salvador; and one of its employees, Ana Perez. Furthermore, my investigation will be aided by publications such as *The Object of Labor: Art, Cloth and Cultural Production* (Livingstone and Ploof 2007) and the *Latin American Fashion Reader* (Root 2003). This methodology and the analysis of my research will hopefully result in a strategy that will dilute the shortcomings of globalization for countries like El Salvador.

As a Salvadorian I find it imperative that these issues be addressed; it is more a matter of equality and fairness than of patriotism.

The Rise and Fall of the Maquiladora

In July 2007 I traveled to El Salvador, with the aim of finding out as much as possible about the infamous maquiladora. Why infamous? The media in El Salvador is always keen on letting Salvadorians know how crooked and corrupt the system is. For instance, Juan Avalos noted on July 24, 2007, in *El Diario de Hoy*, one of the most circulated newspapers in the country, how the financial manager of the maquiladora "Just Garments" had altered the company's financial statements from 2003 to 2006. A foreign investment of US$88,000 could not be accounted for; it later emerged the money had been stolen by company directors.

In a comment posted to the Nicaruaga Extre Blog on February 15, 2006, Gabriel Padilla, a Spanish national volunteering for the less privileged in Nicaragua noted:

The maquiladoras in Nicaragua are places of exploitation where the fundamental rights of employees are not respected; there can be up to 2,500 people working in the factory. The government provides the land so that investment corporations begin business without any benefit to the state in return; salaries are shameful, around thirty cents of a dollar an hour.

The idea of a maquiladora was, initially, to create employment opportunities for a sector that needs it (lower-class women, with little or no education), encourage foreign investment to improve a country's economy and reduce costs for the investors. This is at least what Ricardo Escobar explained to me when we met to discuss the matter in July 2007.

The first *maquila* was established on the Mexican border with the USA in 1964 and the concept quickly began to spread in Central America and some other Latin American countries in the 1970s (Taylor Hansen 2003). In 1961, 4 percent of the clothes sold in the US market were imports. Fifteen years later imports accounted for 31 percent. Today it is approaching 90 percent and still climbing. The increase has mainly come from Third World countries "producing everything from underwear to overcoats" (Howard 2007: 38).

However, in El Salvador the situation is slightly different. Indeed, at one point during the early 1990s there were around 400 maquilas employing 50,000 Salvadorians. Salaries rose from $125 per month to $180 per month in 1995 plus a bonus. With the peace treaty that ended the bloody civil war in 1992, things seemed to be on the rise. According to Escobar (2007), the stability perceived by foreign multinationals was also an incentive for them to invest; I believe that it was purely a wage matter. The facts tell another story. The minimum wage in the USA in the mid-1990s was around $5 an hour, but Howard (2007) argues the cost of labor had been driven so low in the USA that it was competitive to produce there.

So why leave? Unions. By the beginning of the twenty-first century in the USA unions had become very strong, "increasing their wages to $10 an hour" (Howard 2007: 42). In addition, no matter how low the wages in the USA are, the cost of living is lower in El Salvador than it is in the USA—it is the "nature of a Third World country" (Escobar 2007)—so companies can afford to pay their workers less. As a result, the retail price of their products is much less than if produced in the USA, at a 1:15 ratio according to Howard (2007). With the concept of "less is more" (the less you spend the more money you make) in mind, another factor comes into play: Asian economies on the rise, where wages paid to workers are as low as "£1.13 for a nine-hour day" (*The Daily Mail* 2007).

The "Los Capellanos" factory is a small- to medium-scale factory that is owned nationally and manufactures underwear and clothing, mostly for the USA (Figure 1). When speaking to Samuel Cerna, son of the founder and general manager of "Los Capellanos," he was keen to emphasize the impossibility of competing against Chinese and Indian economies:

They expected us from one day to the next to provide the "Full Package", meaning that we had to outsource everything from the buttons to the tags, to the threads to the textiles, all they did was send the design [Figures 2 and 3]. Times changed too fast, we could not adapt. Tell me, how could we jump from having $50,000 to $1,000,000 just like that? Our people could not keep up with the productivity

Figure 1
Underwear construction. The design-plan shows how the pieces need to be cut and arranged, as well as color, threads, and pattern. Maquiladora "Los Capellanos," San Salvador, El Salvador. August 3, 2007. Photo: Carolina Gomez-Aubert.

Figure 2
Product diversification. Shirt manufacturing; once the fabric is cut, it is assembled and sent to the shipping area. Maquiladora "Los Capellanos," San Salvador, El Salvador. August 3, 2007. Photo: Carolina Gomez-Aubert.

Figure 3
Shipping and packaging area. All garments and underwear come here for final inspection, then get packaged and returned to the USA. Maquiladora "Los Capellanos," San Salvador, El Salvador. August 3, 2007. Photo: Carolina Gomez-Aubert.

demanded; we did not have enough employees. So they left. Half of the workforce [50] lost their jobs [Figure 4]. In China or India they can have triple the productivity with a third of the price, obviously they prefer that. (Cerna 2007)

Nevertheless, the Salvadorian media painted a different story. "With the full-package the maquiladora industry will have more advantages to compete against countries like China" (Rodriguez 2003), read the headline of an article published by the

Figure 4
Unproductive resources. Half of the factory floor is empty as a result of cheaper labor in the Asian continent. Maquiladora "Los Capellanos," San Salvador, El Salvador. August 3, 2007. Photo: Carolina Gomez-Aubert.

right-wing newspaper *La Prensa Grafica* in 2003.

ASIC's (Salvadorian Association of Dressmakers Industry) vice-president, Roberto Bonilla, suggests that the biggest challenge for El Salvador is "design, dressmaking and tailoring, for everything else the industry is well prepared" (Rodriguez 2003). Four years after this statement I doubt he would still believe that this holds true. This is where I struggle—if at that point in time there was an awareness of the weaknesses in the industry, why not begin training and teaching people to improve their design and dressmaking skills then? Instead, the summit was aimed at owners and senior executives with conferences highlighting the "monetary growths in the economy" (Rodriguez 2003) that the "Full Package" would bring.

As skeptics predicted, the "Full Package" was a total failure. The president of ASIC, Francisco Escobar Thompson, who is also the owner of INTECU (Textile Industry of Cuscatlan) is an example of this. Until April 2005, INTECU produced clothes for the American company Sarah Lee. On April 12, 2005, the human resources manager informed the Ministry of Employment that due to a lack of orders, 450 employees were to be denied access to the factory as their jobs were effectively terminated until further notice (Ferreira 2005).

The employees were not given any support from the authorities and did not receive any of their salaries owed. In addition, they could not seek work elsewhere as they were still bound by contracts to INTECU. According to Ferreira (2005), the Salvadorian Government estimated "6,000 were unemployed during the first three months of 2005." Escobar (2007) argues that currently that number has gone up to at least 30,000 and there are around 100 maquilas that have survived from the 400 that initially sprawled the country. Adding insult to injury, of these 100 surviving maquilas, three huge companies are responsible for 70 percent of the exports: Primo, Gara, and Fruit of the Loom. The oligopoly gives the false impression of economic growth within the market, when in reality they do not allow anyone else to permeate the so-called "free" market.

Employers wield dictatorial power; massive poverty, unemployment, and no tradition of unionization leaves workers nearly powerless—powerless in monetary terms, powerless in the judicial system, and powerless in their lack of skills and adequate training.

So, what is the next step? I have to considerably agree with Escobar's suggestion (2005) "to transform the maquila industry into a fashion design industry." However, I am more inclined to believe that what could make a substantial difference is craft, which usually takes place in poorer parts of the world. It is initially undertaken out of economic necessity and basic survival, but eventually transitions into sustaining a handmade or traditional practice.

Crossing the Borders

I pondered for many hours over what the basis of the textile industry in El Salvador really is. Gale and Kaur (2002) suggest that the categories or types of textile designers within the fashion and interior design worlds range from "the conceptual to the commercial and the creative to the craftmakers" (Gale and Kaur 2002: 37). In my opinion, it is the level of creativity that varies within those categories that places products higher or lower down in

the consumer's eye. Taking into consideration Gale and Kaur's insight, does El Salvador have that kind of structure?

Designers? Quite unlikely. There are two of them, Jose Dominguez and Jorge Arguett. Both market themselves as *haute couture* designers. In reality, they design evening/party dresses for the high-society women of El Salvador. This extends to the point of going into Dominguez's shop and discovering that half of the dresses there are not even made or designed by him; they are bought in America and sold for double the price as party gowns. On the other hand, Arguett's notoriety relies on his "creation of corsets" (Chavez 2005)—a piece of clothing that has been around for over one hundred years. I do not wish to tarnish their reputation or work because in their own right they have begun to open the doors of fashion design in a country where ten years ago this was unheard of. Fortunately, this statement might have less or no validity in another ten years' time, as USAM (Salvadorian University Alberto Masferrer) in conjunction with the University of North Carolina are establishing "the first textile design course in August 2010" (Escobar 2007).

What about textile art? Gale and Kaur (2002: 77) describe it as a form of art using textiles that can exist in galleries and public spaces, like paintings, sculptures, and installations do. There is one museum of modern art in San Salvador (the capital of El Salvador) where there has never been a fiber art exhibit. "Is that more like craft?" was the response from the museum guide as I explained what

textile art was. As I recall from my visit this past summer, 80 percent of the works were paintings and the rest a mixture of sculptures and installations. I am not suggesting that having a more prominent textile art scene will provide economic relief to the country; however, it may raise El Salvador's profile in a regional or even a global sense, promoting its reputation and bringing the much-needed interest that other areas, like craft, might need in order to thrive.

What about craft? As the museum guide demonstrated, there is a general awareness of craft in El Salvador, and in Latin America generally. Possibly because of its connotations of heritage, process, and technique, being a colonized region places us in a privileged position in the eyes of those outside the borders when it comes to traditional crafts.

In my own practice, I always say to myself "work with what you have," partly because of financial shortcomings but most importantly because I want to be resourceful. So, what does El Salvador have on offer after the maquiladora debacle?

There are 30,000 unemployed people (mostly women), who have skills with industrial machinery that they operated in the maquiladora, knowledge of the components of garments, a visual idea of color and patterns seen from the pieces the American companies send, a need for a job, an instinct for survival, and a untapped creative mind. Moreover, in my research of "Los Capellanos" I found that three-quarters of the people working on the maquiladora floor came from smaller communities outside the city of San Salvador, coming to try their luck into the city. On August 4, 2007, I spoke to Ana Perez, an employee of "Los Capellanos." She comments:

Most of the people in my village Olomega [where she is from] leave to find a better paid job in San Salvador, before I came here I was the seamstress of my community and worked in La Palma, I am good with my hands and had a small stand in the market, I sold clothes that I stitched on and my husband painted wooden crafts. The money was not enough so we had to move to San Salvador, I was pregnant so we needed more money. But I was happier in La Palma ...

There is one phrase that one hears very frequently in reference to La Palma: "Artisan Capital." It began in the 1970s with the arrival of artist Fernando Llort, and his distinctive "La Palma Style." His art style blends "the shapes of Miro, child-like sketches, and mosaic characteristics" (Romero 2005; Figure 5). Drawing upon everyday scenes and nature as inspiration, and his creative discovery of the copinol[1] seed as a material, his style is now recognized throughout El Salvador and beyond (La Palma, El Salvador 2006). There are many artisans who are hugely influenced by him who choose to remain faithful to this style, perfecting details and techniques in the creation of their products (Figure 6).

Except for the products themselves, the style, color, and motifs of Llort's "child-like" designs remain unchanged throughout the years (Figure 7). Why is there no change? This could be in the name of "tradition." Demaray *et al.* (2005:143) define tradition as passing on objects, beliefs or practices unchanged by time. However, they also argue that in order for something to be "traditional" it does not have to be necessarily connected "visually" to the past; it could be "symbolic" (Demaray *et al.* 2005: 159).

In accordance with the economic profile given by La Palma, El Salvador (2006), in 2005 the

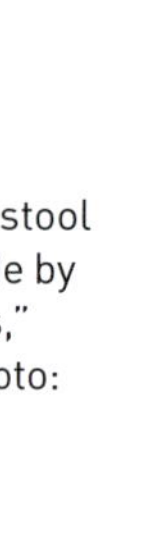

Figure 5
Handmade and painted wooden stool adorned with a farm scene. Made by workshop "El Campesino Crafts," La Palma, El Salvador, 1996. Photo: Carolina Gomez-Aubert.

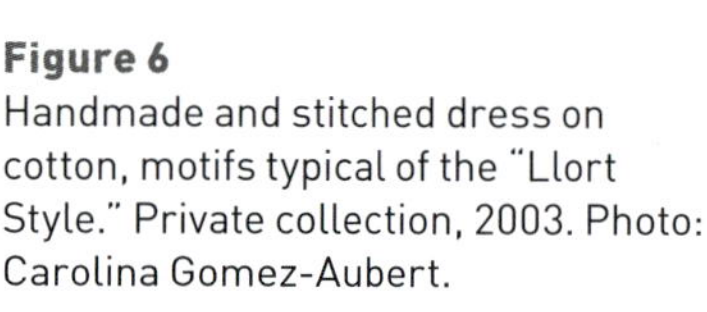

Figure 6
Handmade and stitched dress on cotton, motifs typical of the "Llort Style." Private collection, 2003. Photo: Carolina Gomez-Aubert.

Figure 7
Handmade and painted over-the-shoulder small postman bag. Made by the workshop "Paty," La Palma, El Salvador, 2006. Photo: Carolina Gomez-Aubert.

exportation of artisan products was at its lowest since 1999. There was an oversaturation of the local market with 100 workshops producing almost homogeneous products and 50 percent of them selling within the tourist industry. Gale and Kaur point out that crafts in the modern world are usually seen as an "alternative to mass production" (Gale and Kaur 2002: 70); however, in La Palma the opposite seems to be true—products with Llort's style are mass-produced with little or no individual creativity. I believe this is to blame for the lack of interest in Salvadorian craft that the economic profile shows. There is a need for business refocusing in order to enter or develop a more profitable activity, an activity that reinvents the "La Palma Style" and where these 100 workshops distinguish themselves from one another. As of now one cannot tell whether the products shown in Figures 5–7 came from business

A, B or C, and it seems to me that this is the fundamental flaw of this cooperative.

I have until now ignored another important factor, that Ana Perez reflected on when it comes to people: the employee's self-satisfaction, self-motivation, general contentment, and morale with their job.

For a better understanding of why it is important to fulfill such employee needs I will refer to Maslow's Hierarchy of Needs (*Encyclopaedia Britannica Online* 2007b).

Abraham Maslow developed a model (Figure 8) that suggests once lower-level needs, such as physiological and safety needs, are met then higher-level needs, such as self-fulfillment, are pursued.

Following this theory, then, suggests that people like Ana Perez are far from achieving esteem and self-actualization. In some cases, the situation is more worrying. The following is a description of Corazon Balam's working day in a maquiladora in Merida, Mexico, published on the "Voces de la Maquila" website, February 2004.

We start at 7:15 in the morning and work until 5:30 in the evening; we have 30 minutes for lunch, we don't have time for breakfast, at 3:30 we have 10 minutes break, and if we want to go to the bathroom it has to be fast. Everyone has to take their own scissors, pencil, apron, stitches; they don't even give us breathing protection. There is a lot of pressure and close supervision, the production engineering and the manager watch us all day. We can't talk nor ask for anything...because of this treatment, the women workers only last two or three months. When they see you can do 70% of the production, they start to demand of you 80% or 90%, "You can do three more!" they say...

If Maslow's theory holds, there are important implications for management that should be

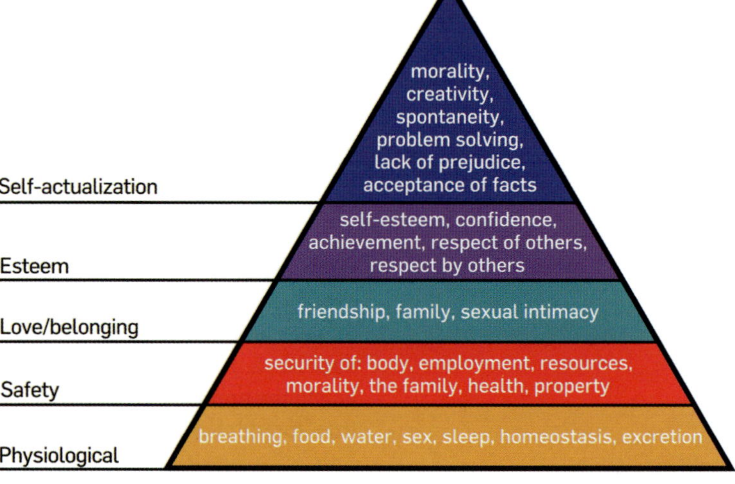

Figure 8
Maslow's Hierarchy of Needs: A pyramid created by Abraham Maslow to explain theories of personality and motivation with primary needs at the bottom. Photo courtesy of FactoryJoe, 2009. Creative Commons Licensing.

considered but are obviously ignored. On its website the Business Knowledge Center (2002) states what each of these needs entail:

- Physiological needs: Providing decent lunch (usually it is an hour) and rest breaks and wages that are sufficient to purchase the essentials of life.
- Safety needs: Such as a safe working environment and job security.
- Social needs: Creating a sense of community via team work and social events.
- Esteem needs: Recognize employee's achievements to make them feel valued.
- Self-actualization: Provide employees with a challenge and the opportunity to reach full career potential.

Both testimonies from Ana Perez and Corazon Mateo demonstrate that these essential needs are barely being met; it really does not surprise me that morale is low. Imagine these two situations multiplied by all the employees of maquiladoras? It would be an error to argue that everyone has the same luck but one person feeling like this is already far too many, especially when there are alternatives.

Coopa Roca is a Brazilian sewing cooperative in a shantytown of Rio de Janeiro that has gone from "a start-up cooperative into a medium-sized national and international operation" (Walbran 2003).

The cooperative that began back in 1981 had six members; it now provides a steady salary for 100 women and temporary positions for at least another fifty. Walbran (2003) explains that initially the cooperative had limited points of sale and their work was considered old-fashioned and was kept at low prices, but it still sold because of its high-quality manufacturing. The founder, Maria Teresa Romeiro, saw lots of potential in the seamstresses' work; it just needed a "fresh twist" (Walbran 2003). Murray (2002) refers to this as the "value chain theory," where a social entrepreneur identifies where and how value can be added to a specific product. Romeiro achieved this by using expensive donated scraps of fabrics and turning them into sophisticated catwalk-worthy garments. Marta Moreiro, a member of Coopa Roca for ten years, says that seeing their work modeled in Milan, London, Berlin, and Sao Paolo gives them huge joy: "that's when we feel we are Coopa Roca" (Walbran 2003). The level of commitment is admirable—ten years of service from the Brazilian cooperative compared to three months from an average maquiladora employee. This is directly related to the level of satisfaction that each of the women feel in their jobs.

Another important aspect about the women in Coopa Roca is that they work at their own pace and handle the load they feel is appropriate to their own skills and levels of productivity. For instance, Marta Moreiro completes thirty-five pieces in eight days (Walbran 2003) whilst in "Los Capellanos" they are expected to assemble at 100 percent productivity 757 pieces in twelve hours (one working day)! Of course the nature of the work differs. At Coopa Roca, Romeiro explains they do not "churn out bric a brac, there is original craftsmanship and the use of finest materials to create big ticket items" (Murray 2002), while in "Los Capellanos" the aim is to produce as much as possible and if you have a surplus of 115 percent then you get a bonus of $6.52 on top of your monthly salary of $150 (Cerna 2007). The members of Coopa Roca can earn up to $263 per month (Walbran 2003) depending on their chosen workload. In addition, these women have the opportunity to work at home, which means they can look after their children without having the added expense of childcare.

If, for instance, "Los Capellanos" became a cooperative where the aim was to produce clothing that was partially or totally crafted, sophisticated, and beautiful, where would they be? Motivated, earning better salaries, and possibly seeing their work in the catwalks of Paris, London, and Milan! Even in terms of technology, which Walbran (2003) cites as "essential" in developing a business of this sort, they have an advantage because they know how to use industrial machinery, unlike in Rio de Janeiro's Coopa Roca where some had never even "turned a sewing machine on" (Walbran 2003).

Overall, Coopa Roca gives a sense of community to the workers and they regard the community as a "second family" (Becker 2005). Furthermore, their reputation has grown so much that they are now working with renowned designers around the world, such as Carlos Miele, Christian Lacroix, Tord Boontje, and artists such as Ernesto Neto. Future plans even involve a collaboration with corporate giant Puma. Essentially, from handcrafts they have crossed over into high-end fashion and blurred the boundaries into textile art. It is the artists' and designers' vision, but it is Coopa Roca's hands that bring them to life.

With the Coopa Roca model there are certain loopholes that need to be considered in order to even begin to suggest that a model like this one can work within the Salvadorian community.

Firstly, the coordinator of such a project must have contacts with "high culture and its money" (Becker 2005) to prevent it from becoming a Third World folklore niche. It is extremely unlikely that ex-employees of maquiladoras would have such contacts. Unlike Rio de Janeiro's international fashion scene, El Salvador does not have one, yet. But, once this scene begins to develop, it could help create the demand that can sustain an avant-garde craft industry. Globally, there is a trend where consumers expect producers to consider aspects of "ethical behaviour and ecological policies" (Gale and Kaur 2004: 86); I do not think this vision has reached El Salvador entirely, but advances in communication and technology shrink time and space allowing the trend to arrive at a faster pace than, say, ten years ago.

There is another fear—that this sort of production in the end will result in "products made using the same technique like from an assembly line" (Becker 2005). This, to me, is more a matter of choosing carefully whom to collaborate with and how long for. As had happened with the maquiladoras, where three major companies were absorbing most of the factories, the same occurred with Coopa Roca.

A large-scale order from Carlos Miele brought a lot of work but they did not want to just work on crochet for ten years, so they ended the cooperation and eliminated that "sense of dependence that a single relationship alone can bring" (Walbran 2003). This way there is not only a sense of control on behalf of the cooperative, but there is also strategic development decision to be taken as you would in any other business.

"Craft is in the throes of a 21st century renaissance," commented Louise Taylor, the UK Craft Council director, in 2003 (Bunnell 2004). I must say that for me this renaissance is closely related to innovations that combine two fundamental concepts: cooperatives and product differentiation aided by technology. The more sophisticated the transformation process, the more value is created in the final product. Is a craft approach, though, solely the answer?

Balancing the Unbalanced

It was only recently that I heard the following statement:

> *What is wrong with mass production? What is wrong with an exquisite one-off piece? Are they not mutually exclusive? One needs the other to subsist. That is the only way we can appreciate their real value.*
> (Siddiqui-Lester 2007)

Personally, I had always shrugged my shoulders at the idea of mass production, classifying it as boring, with no creative human input (except in the original prototype), and horrid conditions for labor—all in all just a negative industry to support. The truth is, looking around me, 95 percent of my surroundings are mass-produced: Excluding a Salvadorian woven fabric on the couch, a hand-painted Greek plate by the window, and some of my studio work on the table.

As Siddiqui-Lester (2007) suggests, it is this imbalance that makes me value these latter works higher than, for instance, my laptop. If this is true, then it would be fair to put forward the idea of a combination between the mass-produced and the more sophisticated one-offs, in order to achieve an even higher value (imagine a laptop with a fine handmade keyboard of delicate silk threads, instead of plastic), not only in monetary terms, but also the value associated with the labor that goes into producing it.

Could mass-produced commodities with handmade embellishments be a solution to El Salvador's crisis?

"Never," says Cerna (2007) with a very definite tone in his voice. With a bit of resentment for his lack of belief, I ask why and he answers:

There is too much bootlegging happening, thieves mug lorries carrying the merchandise and then sell everything so cheap in the black market, remember that for them everything is a profit since it is stolen. The other problem is that people would rather wait for their relatives to send them money from up north [meaning the USA], than have to work in crafts or anything for that matter ... (Cerna 2007)

He then proceeds to recount the story of how he got a bottle of Chilean red wine, Casillero del Diablo, worth $25 in the supermarket, for $2 on the black market, courtesy of one of his employees.

Both are very valid points; however, I have to say that as long as there is a demand there will be a supply. I understand how tempting it is to get something that is particularly expensive for a bargain price, but by doing so we are worsening the situation for thousands of people. I do not wish to condemn Cerna for his actions because he is not the only one, but if he and the thousands of people who fall into this would avoid buying from the black market then just maybe Salvadorians can begin to climb up the "downward spiral" (Howard 2007: 43) that such countries like El Salvador are well known for.

The issue of external monetary consignments is a more serious one. In a comment posted to the jjdalton blog on December 19, 2006, Juan Jose Dalton[2] noted monetary consignments accounted for "17% of El Salvador's Gross Domestic Product (GDP) in 2006," with an incoming flux of three thousand million US dollars. According to Dalton, the monthly consignment sent to a family member from another in the USA is around $250. This figure is almost double the regular wage in a maquiladora and even a minimum wage anywhere else, so Cerna validly points out that people would rather have that income than work. On its website USAID points out that 1.2 million Salvadorians live in the USA and most of them illegally; they suggest that this number has dramatically increased in the past three years especially in the male population. This is where government entities need to intervene. What happens to a family's income if the husband[3] that went illegally to the USA is deported? What happens if the "husband" ceases consignments; for instance, he met another woman, laws getting stricter for immigrants, etc.? On the roads of San Salvador I see advertisements for businesses such as Western Union[4] everywhere, but I have never seen one billboard that says "Women: Help Yourselves, Help your Country." I have never seen television advertisements that explain what would happen if on top of receiving a consignment from abroad, women worked to earn their own salaries? What satisfaction could this bring to the female population? How could this help the economy? It is the government's duty to encourage this sort of thinking, instead of promoting the fact that now that our local currency is the US dollar "the economy has grown at an annual rate of 2%, reaching a 2.8% in 2005" (USAID 2006). Ironically, according to its website, ECLAC (The Economic Commission for Latin America, a non-governmental entity) suggests El Salvador is the country that "grows the least and receives the least amount of foreign investment" (ECLAC 2005).

Project Alabama began in 2001 using labor from low-income families in Florence, Alabama. The idea behind the label was to keep everything sourced and made in the USA; by 2005 they were selling in sixty stores worldwide. However, production came to a halt on September 2006 (Tang 2007).

What happened? On the one hand, Tang (2007) suggests that there was no strategy when utilizing native work and delivering on time. Enrico Marone-Cinzano, co-founder of the company, agrees: "part of the problem was the company's manufacturing operations were too time consuming" (Tang 2007: 22). By contrast, Poko Pano, a Brazilian swimwear company, produces hundreds of bikinis a year, but there is "only one that is completely hand knit lace—for the rest it is an embellishment, this is how we can plan for the ability to meet the demand" (Snyder 2006). Poko Pano's method of production is then a balanced combination that allows high volumes of production, but at the same time they help the local economy, as it is women within the factory that get hired to do the handmade embellishments.

If "Los Capellanos," for instance, were to reinvent their methods of production maybe this is a model they could adopt. In this manner it will not be an abrupt change but a slow transition that mixes assembly-line manufacturing with handmade elements. In addition, the handmade elements would increase the value of the product and seamstresses may begin to feel that their small but vital contribution really makes a difference within the company, increasing self-motivation and satisfaction, which as Maslow's theory points out, is essential for a person's achievement of self-actualization. Furthermore, if these handmade elements were to be used season after season then they can rely on the income and the job, creating the stability that cannot come from relying on consignments from abroad.

Another issue Cinzano (Tang 2007) points out is the difficulty of managing 150 seamstresses working from home, dealing with their personal issues, and all that labor which drives up the retail price. Firstly, this might boil down to a cultural issue. At Coopa Roca, there are around 100 seamstresses plus fifty temporary women, and working at home is actually considered an advantage because they are able to look after their children as well. I believe it might have a bit to do with the following statistic: the Inter-Union Department of Statistics and Socio Economic Studies shows that the entrepreneurial spirit in Brazil is a lot higher than in the USA. In Brazil one in five runs their own business, in the USA it is one in ten, whilst in Japan it is one in 100 (Walbran 2003). In Latin American culture, especially if you come from a low-income family, it is extremely important to be an exceptional housewife and if this can be combined with employment that is really lucky.

Secondly, as mentioned before, labor in the USA is paid at a higher rate than it is in Latin America so it is only natural that the retail price of the final product would be a lot higher. It is at this point that Ying Li, a designer of handpainted lingerie, chooses to outsource to China, as she comments, "if something is done better in China, go to China. It is a trained artisan who knows how to paint in silk" (Tang 2007: 23). This, however, defeats the whole purpose of the vertical production concept that Project Alabama was developing, the same concept El Salvador could greatly benefit from. Supervising the quality

Figure 9
Alabama Chanin, Reverse Appliqué Technique. Layers of cotton jersey joined by quilting then the top layer is cut away to reveal color underneath. Work presented at the *Ever and Again* Exhibition, Triangle Gallery, Chelsea College of Art and Design, London, October 2007. Photo: Carolina Gomez-Aubert.

of the product must be almost impossible if your work is half way around the world and there is a high margin of error as well, and let us not forget the environmental impact of transportation. If you can create a community within the company then you can build what Cinzano (Tang 2007: 23) refers to as the "pyramid effect," where one person above helps the person below. Precisely what Coopa Roca has developed—Maria Macedo Pereira's previous business was not running well, so she took a course on a certain technique at Coopa Roca, passed the test and today trains young women in the same technique (Becker 2005). Fortunately, for a country like El Salvador, where labor is as cheap as it is in Brazil, the likelihood of a model like this working is higher than it is in the USA. And, if the entrepreneurial spirit is as high as it is in Brazil, opportunities begin to be viable ones.

After all the problems Project Alabama encountered, co-founder Natalie Chanin created the spin-off company Alabama Chanin. She now heads a smaller community of fifteen people, who create limited-edition jewelry, clothing (see Figure 9), and home furnishings, made in Alabama with organic and recycled materials. There is a controlled level of production and "seamstresses do not work at home, they attend a production centre 8 hours per day" (Shi 2006). Coincidentally enough, Chanin graduated from North Carolina University, the establishment that, collaborating with USAM, will begin the first textile design program in El Salvador.

This sort of smaller-scale, limited-edition production applies better to the small workshops in La Palma. Due to the smaller number of employees within the businesses they could make products that still reflect the Llort style but with a twist, something that develops the already existing and over-used techniques of shape, color, and materials; as Wagner (2007) puts it, "modernised old techniques."

A change of this sort needs to be guided by someone who has a vision, just as Natalie Chanin did for Alabama Chanin and Maria Teresa Romeiro did for Coopa Roca; it is an optimal opportunity for the new generation of social entrepreneurs, like myself, to incorporate innovative schemes of design and development to keep the La Palma type cooperatives alive, and give birth to new ones.

Conclusion

El Salvador is indeed a new gem; it is a country that had to go through a bloody civil war for over ten years in order to see the first major change in its political scene. Fifteen years have passed since it ended; not enough time to rebuild a distraught society but enough to open up new doors.

After careful consideration and analysis the future is not a bleak one for El Salvador's textile industry; the possibility of developing and sustaining homegrown design is a palpable one. It is a combination of elements discovered during this investigation that could allow this.

Firstly, the fact that the first textile design program will begin in August 2010 is a vital piece of the puzzle. According to Escobar (2007) people will be receiving adequate training and teaching in areas such as technology, branding/marketing, pattern cutting, and design. This will, in turn, deliver educated individuals that will allow employers to compete in a more stable industry on the "basis of technological innovation, superior design and more effective marketing" (Howard 2007: 50). In addition, having the support of a recognized institution in North Carolina offers USAM credibility not only nationally but also internationally.

Nevertheless, I have to admit this does not mean that the maquiladora will cease to exist. Although the number of maquiladoras in the country has decreased by three-quarters and Escobar (2007) suggests they will become "taboo," the remaining ones should begin to consider the incorporation of mass-produced commodities with hand embellishments—a scheme that has been successful for Poko Pano in the last twelve years. The integration of handcrafts in a maquiladora will empower its employees, increasing their motivation and job security as well as their wages. Moreover, they should unionize if there is any discontent in their working environments, disrupting production if necessary. It is not solely a matter for the employees; employers that resist the establishment of decent standards should be penalized through government enforcement and by consumers' rejection of goods produced in an insensitive manner. All of these become weapons for both consumer and worker and, as my economics teacher used to say, "any negative word of mouth or publicity will be passed on to ten people, whereas anything positive only to one."

Secondly, the creation of cooperatives whose products are non-homogenous is an essential factor for the country to develop and sustain a homegrown design industry. Although these already exist in La Palma there is no product differentiation; this can change with the aid not only of students who study abroad but also by the future generations of USAM graduates. The craft industry has then got the building blocks to shed their folksy image and enter the market of highly desirable, consumer goods, gaining greater respect in the design and art worlds, something that Coopa Roca began to achieve over twenty years ago. In addition, if consumers are satisfied with the product, premiums can begin to be introduced where goods are certified for meeting certain standards, stimulating growth for the small producer organizations. I am not suggesting that all this will occur in, say, two years' time. On the contrary, it will take time, but if it were applied it is a long-term plan that would see its rewards in, say, twenty years' time.

Finally, the people employed within the industry are mostly women and it is this sector of society where efforts need to be aimed. The twenty-first century is providing the stage for this. In 2008, Hillary Clinton and Sarah Palin were both running for top positions in the US government. In El Salvador we seem to absorb quite easily anything that originates from the USA, from their currency to supporting the war in Iraq;

this is an excellent opportunity to focus on a more beneficial aspect: the exploitation of women in leadership roles. As a result, this could inspire Salvadorian women to make a difference not only in their individual households, but also for the entire country.

In the essay "Fashioning United States Salvadoranness" Milian Arias claims that El Salvador's society is "absent of reason" (2005: 271). I believe that it has been for quite some time, especially in the current political climate. This is where the unreasonable part of society lies. It is this segment that must rationalize and be humanized because it is only when we begin to work together, rather than against each other, that the pieces of Salvadorian design will fall into place.

Notes

1. An oval seed, often split, from a Salvadorian tree.
2. He is a noted political Salvadorian journalist, international issues correspondent, and director of *ContraPunto* magazine.
3. The sole or main income earner of a typical Salvadorian household tends to be the husband.
4. Consumer-to-consumer money transfer services to get resources to family members around the world. They have over 370,000 agent locations in over 200 countries.

References

Becker, Jochen. 2005. "Every Ground Floor is Converted into a Workshop." City of Coop. http://www.metrozones.info/metrobuecher/coop/becker_02.html (accessed November 2, 2007).

Bunnell, Katie. 2004. "Craft and Digital Technology." Paper presented at the World's Craft Council 40th Anniversary Conference, June 2, Metsovo, Greece.

Business Knowledge Center. 2002. "Maslow's Hierarchy of Needs." http://www.netmba.com/mgmt/ob/motivation/maslow/ (accessed August 25, 2007).

Cerna, Samuel. 2007. Interview with General Manager of "Los Capellanos." San Salvador, El Salvador, August 4.

Chavez, Glenda (trans.). 2005. Interview: Jorge Arguett "My Inspiration is Women." http://www.elsalvador.com/mujeres/2005/12/entrevista/index.asp (accessed October 20, 2007).

Demaray, Elyse, Keim-Shek, Melody and Littrel, Mary A. 2005. "Representations of Tradition in Latin American Boundary Art." In Regina Root (ed.) *The Latin American Fashion Reader*, pp. 142–59. Oxford: Berg Publishers.

ECLAC. 2005. "Central American Region: Economic Evolution in 2004 and 2005." http://www.eclac.cl/cgi-bin/getProd.asp?xml=/publicaciones/xml/0/20870/P20870.xml&xsl=/mexico/tpl/p9f.xsl&base=/mexico/tpl/top-bottom.xsl (accessed November 23, 2007).

Encyclopaedia Britannica Online. 2007a. "Maquiladora." http://en.wikipedia.org/wiki/Maquila (accessed October 5, 2007).

Encyclopaedia Britannica Online. 2007b. "Maslow's Hierarchy of Needs." http://en.wikipedia.org/wiki/Maslows_hierarchy_of_needs (accessed October 28, 2007).

Encyclopaedia Britannica Online. 2007c. "GDP." http://www.britannica.com/EBchecked/topic/246647/gross-domestic-product-GDP (accessed July 14, 2011).

Escobar, Ricardo. 2007. Interview with former owner of a maquiladora, current member of ASIC. San Salvador, El Salvador, July 20.

Ferreira, Walter (trans.). 2005. "Law of the Maquila, Law of the Jungle: There Is Progress That Kills." June 24. http://www.rel-uita.org/sindicatos/maquilas/ley-maquilas-1.htm (accessed October 14, 2007).

Gale, Colin and Kaur, Jasbir. 2002. *The Textile Book.* Oxford: Berg Publishers.

Gale, Colin and Kaur, Jasbir. 2004. *Fashion and Textiles: An Overview.* Oxford: Berg Publishers.

Howard, Alan. 2007. "Labor, History and Sweatshops in the New Global Economy." In Joan Livingstone and John Ploof (eds) *The Object of Labor: Art, Cloth and Cultural Production*, pp. 31–50. Chicago, IL: School of the Art Institute of Chicago Press.

La Palma, El Salvador. 2006. "La Palma's Artisans: Economic Profile." http://www.lapalmaelsalvador.com/en/aboutart/econ.shtm (accessed October 19, 2007).

Livingstone, Joan and Ploof, John (eds). 2007. *The Object of Labor:*

Art, Cloth and Cultural Production. Chicago, IL: School of the Art Institute Chicago Press.

Milian Arias, Claudia M. 2005. "Fashioning United States Salvadoranness: Unveiling the Faces of Christy Turlington and Rosa Lopez." In Regina Root (ed.) *The Latin American Fashion Reader*, pp. 263–79. Oxford: Berg Publishers.

Murray, Inez. 2002. "Create Value and Create it NOW!" *Changemakers Journal* June. http://proxied. changemakers.net/journal/02june/ murray.cfm#jump (accessed October 28, 2007)

Ricardo, David. 2006. *On the Principles of Political Economy and Taxation.* New York, Cosimo Inc. Publishers.

Rodriguez, Gustavo (trans.). 2003. "Forward with Central American Summit on Maquilas, La Prensa Grafica." http://archive.laprensa. com.sv/20021206/economia/ ecm11.asp (accessed October 13, 2007).

Romero, Tiana. 2005. "El Salvador's Celebrator of Life." Ella House Fine Art Gallery. http://www.ellahouse. com/Education/miscellaneous/ celebrator_of_life.php (accessed October 25, 2007).

Root, Regina (ed.). 2005. *The Latin American Fashion Reader.* Oxford: Berg Publishers.

Shi, Jim. 2006. "Sweet Home: Alabama Chanin Gets Its Lifestyle on." *Fashion Week Daily* November 28. http://www. fashionweekdaily.com/news/ fullstory.sps?inewsid=382846 (accessed November 14, 2007).

Siddiqui-Lester, Zara. 2007. Business Start-Up Seminar. Chelsea College of Art and Design, London, November 13.

Snyder, Amy. 2006. "Poko Pano Swimwear." *Cashak Magazine* June 23. http://www.cashakmagazine. com/interviews/2006/poko-pano- amy-snyder/ (accessed November 13, 2007).

SubRosa. 2007. "Can You See Us Now? Ya nos Pueden Ver?" In Joan Livingstone and John Ploof (eds) *The Object of Labor: Art, Cloth and Cultural Production,* pp. 80–8. Chicago, IL: School of the Art Institute Chicago Press.

Tang, Syl. 2007. "The Hand's Made Tale." *The Financial Times*: The Business of Fashion, Spring: 22–3.

Taylor Hansen, Lawrence. 2003. "The Origins of the Maquila Industry in Mexico." *Bancomext Magazine* 53 (November 30). http:// revistas.bancomext.gob.mx/rce/ magazines_en/24/6/tayl1103.pdf (accessed February 4, 2008).

The Daily Mail. 2007. "The High Street Sweatshops: Primark and M&S Factories Pay 13p an Hour." *The Daily Mail* September 3. http:// www.dailymail.co.uk/pages/live/ articles/news/news.html?in_ article_id=479571&in_page_id=1770 (accessed October 25, 2007).

USAID. 2006. "Country Overview." http://www.usaid.gov/sv/country. html (accessed November 7, 2007).

Walbran, Shannon. 2003. "Shanty Town Seamstresses Fuel the Fashion Industry." *Global Envision* May 20. http://www. globalenvision.org/library/10/477 (accessed August 23, 2007).

Touching the Hem: The Thread between Garment and Blood in the Story of the Woman with the Haemorrhage (Mark 5:24b–34parr)

Abstract

In this article I start from a particular passage in the New Testament that will touch on biblical threads and paradigms related to textile. Mark 5:24b–34, Luke 8:42–48, and Matthew 9:19–22 tell the story of the healing of the "woman with an issue of blood," the Haemorrhoissa. This story (and its *Nachleben* in commentaries and iconographies) pulsates with a delicate energy relating to textile, cloth, and the magical impact of touch. I will firstly investigate upon those textile-related dimensions by situating the passage in what I term its narrative, iconic, and anthropological space. This will finally lead to a hypothesis on the origin of the medieval assimilation of the Haemorrhoissa to the Veronica, the woman who carried the imprint of Christ's face.

Keywords: Haemorrhaging Woman, Christianity, textile, cloth, hem, visuality, *vera icon*

BARBARA BAERT (WITH THE COLLABORATION OF EMMA SIDGWICK)

Barbara Baert is professor of art and art history at K. U. Leuven. In 2008 she founded the Iconology Research Group, an international and interdisciplinary platform for the study of the interpretation of images. Her recent books (edited) are *Weaving, Veiling, and Dressing: Textiles and Their Metaphors in the Late Middle Ages* (Brepols 2007) and *Fluid Flesh: The Body, Religion and the Visual Arts* (Lieven Gevaert Studies in Photography and Visual studies 2009).

Textile, Volume 9, Issue 3, pp. 308–351
DOI: 10.2752/175183511X13173703491117
Reprints available directly from the Publishers.
Photocopying permitted by licence only.
© 2011 Berg. Printed in the United Kingdom.

Touching the Hem: The Thread between Garment and Blood in the Story of the Woman with the Haemorrhage (Mark 5:24b–34parr)

*Il n'y a pas d'abord la
    signification, la traduction,
    l'interprétation:
il y a cette limite, ce bord, ce
    contour, cette extrémité, ce
    plan "d'esposition",
cette couleur-sujet locale,
qui peut simultanément se
    contracter.(...)
Cela seulement peut fermer
    ou dégager de l'espace
    pour des "interprétations."*
(Nancy 2000: 23)

*At the outset, there is no
    signification, translation, or
    interpretation:
there is this limit, this edge, this
    contour, this extremity, this
    outline,
this local subject-color,
which can be withdrawn,
    concentrated, (...).
This alone can close or release a
    space for "interpretations."*
(Nancy 2008: 23)

One of God's first acts of provision for humankind after the fall is to make "garments of skins for the man and for his wife and clothe them" (Gen. 3:21) (Rubin and Kosman 1997). The world of the Bible, especially for what concerns its Christian understanding, is covered with garments, wool, cloths, textiles, threads, and veils. Not speaking and writing, but knotting and weaving seem to be the appropriate ways to understand Christian anthropology. The seedbed of this Christian anthropology can indeed be found in the Old Testament; already their clothing is rarely neutral (Ryken 1998: 318–9). As we understand from the prophets, it is the image for those who live in the presence of the Lord (Ezekiel 16:13 and Isaiah 23:18). Nature is part of the clothing of God, wrapped in light as a garment, sings Psalm 104:2. There were garments unique to prophets, such as Elijah's mantle (1 Kings 19:13). Putting on and taking off clothes signifies the transformation from one role or state of mind to the other, such as priesthood, mourning or becoming a bride. Garments help to link the natural and supernatural world. The weaving of the thread in the womb of Psalm 139:15 ("You knit me together in my mother's womb"), and Moses veiling his face after having seen God in Exodus 34:28–35, preserved the very seed for the fundamental paradigms of incarnation and visible invisibility. In fact, garments and by extension textiles are mysterious and multi-sensory threads connecting the glory of God to mankind that is at His service.

In this article I start from a particular passage in the New Testament that will touch these multi-sensory threads and paradigms. Mark 5:24b–34, Luke 8:42–48, and Matthew 9:19–22 tell the story of the healing of the "woman with an issue of blood," the Haemorrhoissa.[1] This story (and its *Nachleben* in commentaries and iconographies) pulsates with a delicate energy relating to textile, cloth and the magical impact of touch.

And a large crowd followed him and pressed in on him. [25]*Now there was a woman who had been suffering from hemorrhages for twelve years.* [26]*She had endured much under many physicians, and had spent all that she had; and she was no better, but rather grew worse.* [27]*She had heard about Jesus, and came up behind him in the crowd and touched his cloak,* [28]*for she said "If I but touch his clothes, I will be made well."* [29]*Immediately her haemorrhage stopped; and she felt in her body that she was healed of her disease.*

[30]*Immediately aware that power had gone forth from him, Jesus turned about in the crowd and said, "Who touched my clothes?"* [31]*And his disciples said to him, "You see the crowd pressing in on you; how can you say, 'Who touched me?'"* [32]*He looked all round to see who had done it.* [33]*But the woman, knowing what had happened to her, came in fear and trembling,* *fell down before him, and told him the whole truth.* [34]*He said to her, "Daughter, your faith has made you well; go in peace, and be healed of your disease."* (Mark 5:24b–34)[2]

The Narrative Space: Touch and Textile

The Haemorrhoissa is the interruption of Jesus' requested cure of the daughter of Jairus. The bleeding woman appears suddenly out of the crowd with a strong desire to be cured of her hemorrhage—*siccatus est fons sanguinis eius*. The reference to the *fons* in verse 29, refers to the fact that her hemorrhage was a uterine problem, and not for example a nosebleed. Another argument in the secondary exegetical literature, is that the fact that the flow not being located exactly—*mulier quae erat in profluvio sanguinis annis duodecim* (Mk 5:25)—might indicate discretion, even shame in Mark's redaction. Marla Selvidge and Judith Plaskow discern anti-Judaic sympathies in Mark, in view of a deliberate healing of a woman that would be an outcast in Jewish society,[3] given the purity regulations in Leviticus (12:7, 15:19–33, 20:18), Numbers and Deuteronomy (Plaskow 1993; Selvidge 1990).[4] The woman with the hemorrhage might have been a so-called *zabâ*, a woman suffering from irregular menstrual bleeding; one who, according to Leviticus, could not be touched.[5] In short, the passage in Mark becomes contextualized in paleo-Christian culture, in menstruations and blood-flow taboos with a central focus on tactility.

Christ's miracles of healing are carried out through the word, through touch, through word and touch, with the use of saliva, at a distance, or through incision (van der Loos 1965: 510; Wilkinson 1998). The Haemorrhoissa clearly belongs to the category of touch: *haptein, tenere*.[6] But the touch of the woman with the hemorrhage is peculiar for three reasons. To begin with, she initiates the touch, and the touch seems to have a strong effect on Jesus. Her fountain of blood dries out, at the exact moment that something "flows away" from Jesus. "This was no accidental brush with a stranger; it was a risky endeavour to make physical contact with the person of Jesus, if even his garment" (Bromley 2005).

Reimund Bieringer (2005) has studied the syntactic meaning and frequency of the word "touch" in the Old and New Testament. The Greek verb *haptein* is the most general verb for touching, but also means "to approach," "to be in contact with something or someone" or "to touch emotionally" (both in a friendly and in a harmful sense). Comparative research of the frequency and the contextual meaning of the verb *haptein* has shown a cultic meaning (Ex 29:37) or a taboo of touch (Leviticus and Numbers) between people, things and dead bodies. The same word is actually used in the *Noli me tangere* phrase: *me mou haptou* (John 20:17).[7] In his article *Healing by a Mere Touch*, Lalleman states that healing by touch is less typical of Hellenistic cults than of Judaeo-Christian healing practices (1997).[8] Greek and Roman healing practice appealed to therapy, such as the bath therapies and dreams of the Asklepios cults (Cotter 1999: 246).[9] Mark and Luke specify, moreover, that therapists had been unable to help the woman. Christ's healing, by contrast, takes place in a flash, a *momentum*, with no means employed, no process or therapy, and thus uniquely connects the touch—*haptomai*—with the possibility of healing in a miraculous way.[10]

Secondly, the touch causes a fascinating breach between fluidity and petrifaction, between movement and fossilization, between process and interruption. The text even speaks of a power (*dynamij*; *virtus*) over which Christ has no control. One might sense an archaic magical effect suggested here (Bromley 2005; Clark Kee 1983: 162–3, 1984: 2–4; Cotter 1999: 246; Hull 1974: 136). According to Hippocrates (460–377 BC) *dynamij* means the capacity of affects: a mana-like fluid. It is an energy that is transmitted from the supra-mundane to the mundane level. Normally *dynamij* is not used in the Christian miracle episodes. It refers to a Hellenistic cosmic view of healing.[11] But if the synoptics allude to old magical suggestions, then it is to thematize their very denial by Christ (Marshall 1989: 106–8; May 1952: 98). At the end of the episode the healing is attributed to Christ's person. He explains the miracle by referring to the woman's faith: "your faith has made you well."

Finally, we need to mention that this touch is a touch upon textile. It is, in other words, no skin-to-skin contact such has the typical laying on of hands (*sunanaxroosis*)

(Waagenvoort 1957). The idea that power can be stored or tapped from a holy person is very old (Marcus 2000: 359). In 2 Kings 13:20–21 a strong example of this is preserved: a corpse that touches the bones in Elishas's tomb revives. Also textile has the explicit capacity to transfer essences of the owner's body. The clothing of a body constitutes that body. This is also clear in Jesus asking, "Who touched my clothes?" and the disciples' reaction: "Why would you ask 'Who touched me?'" In Acts 19:12, handkerchiefs that have touched Paul's body have the power to heal the people of Ephesus. In the Acts of John, many brothers were healed in Ephesus by touching the garments of John; the word used here is, indeed, *haptomai*.[12]

Again, according to Leviticus 15:11, 21–22, one needs to wash not only the hands but also the clothes of one who has been touched by the afflicted. Touching a menstruant, in particular, can cancel miraculous power ("a power had gone from me"). In the Jewish Hekhalot Rabbati of the third–fourth century, a Rabbi lost his power by touching a thread that had been touched by a menstruant (Marcus 2000: 358). There are also Greco-Roman examples, such as Epidaurus 7, which states that a disease can not only be transmitted to a cloth, but be transferred in full from the sufferer to the cloth (Theissen 1983: 134). Some authors take the view that the hemorrhaging woman wished to get rid of her decease by passing it on to the garments of Christ (Theissen 1983: 134). Whether this is the case or not, the episode of the Haemorrhoissa shows some sort of comparable superstition at work, with the idea that touch can draw out a power. Mark's agenda was probably to show that faith can overcome superstitious fear (D'Angelo 1999: 99).

It is important to understand that for all these reasons the touch by the woman with the hemorrhage is different from other biblical healing stories. In the next section, I will investigate these characteristics in early medieval visual culture. In other words, we will make the transposition from the narrative space to the iconic space.

The Iconic Space: "Textilization"

Asterius, Bishop of Pontus in Asia Minor, wrote *c.* 400:

> *The more religious among rich men and women, having picked out the story of the Gospels, have handed it over to the weavers—I mean our Christ together with all His disciples, and each one of the miracles the way it is related. You may see the wedding of Galilee with the water jars, the paralytic carrying his bed on his shoulder, the blind man healed by means of clay, the woman with an issue of blood seizing Christ's hem, the sinful woman falling at the feet of Jesus, Lazarus coming back to life from his tomb. In doing this, they consider themselves to be religious and to be wearing clothes that are agreeable to God.* (Asterius of Amasia 1972: 51)

The iconography of the Haemorrhoissa was a frequent theme in paleo-Christian visual culture appearing on sarcophagi, in catacombs, and on small objects such as textiles, ivories and gems. Already in the catacomb of Peter and Marcellinus alone, five different identifications of the Haemorrhoissa are accepted in the secondary literature (see Deckers *et al.* 1987; Mancinelli 1987: 39–43). The fact that the theme is so popular in this catacomb makes some scholars presume that all women were buried here (Mancinelli 1987). It is moreover the catacomb also known as "ad S. Helenam," referring to the mausoleum (on the same site) of the first Christian empress, the mother of Constantine the Great.

In his essay *À distance* Carlo Ginzberg connects the idea of the *Punctum* or *l'instant decisive* with the very birth of Christian art.[13] The *Zwischenraum* between the biblical text and the emergence of an iconography is characterized by vibrant moments that have the capacity to open the narrative space to the iconic. The text breaks open towards the epiphany of the visual. We can discern three such pulsating moments in the mother-text (see Baert 2009a).

The first group refers to the crawling woman, the outcast, who desires to touch (Figure 1). Tactile contact is either taking place or about to take place. The gazes between the woman and Christ do not, however, meet, as Christ is still searching elsewhere. Nor do their bodies connect. The distortion—she in profile, he frontal or turning around—embodies the transgressions that are taking place. The second group is the woman falling down, trembling, and explaining her actions in anxiety (Figure 2). Christ is looking down at her. She has lost her grasp, and fears punishment: *venit et procidit* (v. 33). The third group refers to the harmony and final stage of the story (Figure 3). The woman is healed, blessed, and become "whole." The woman is semi-upraised, accepting the blessing and final Christological healing moment: *fides tua te salvam fecit* (v. 35). The transgressing energies give way to stillness and union. There is no longer any tactile contact with Christ. The wholeness is also expressed in the mutual gazes passing between Christ and the woman. The final connection is made.

But in none of these cases is the pathology of the woman with the hemorrhage recognizable. In the images of the healing of the blind man at Siloam (Jn 9:1–4) (Baert 2007) the eyes are pointed to, and in those of the paralytic at the *Piscina Probatica* (Jn 5:1–8) the bed or the crutch refer to the specific handicap (Baert 2005). But in the case of the Haemorrhoissa nothing allows us to induce that the miracle might involve a hemorrhage of the uterus. It is undepictable. This can sometimes make her difficult to identify; she is often confused with Mary Magdalene, the woman of Canaan (Mt 15:22) and the humpbacked woman (Lk 13:10–17).[14] In the miniature of the *Codex Egberti*, the miniaturist added above the woman: *fluxus habens* (Figure 4).[15]

In the Gospels, the illness of the woman is compared to a *fons sanguinis*. When one considers the *passus* as an image of the well, it suddenly becomes apparent how

Figure 1
Christ and the Haemorrhoissa, detail of
the *Brescia Casket*, ivory relief,
c. 360–70. Brescia, Museo Civico.

Figure 2
Christ and the Haemorrhoissa, mosaic,
520–26. Ravenna, S. Apollinare Nuovo.

Figure 3
Christ and the Haemorrhoissa, mural in the catacomb of SS Peter and Marcellinus, third century. Rome.

Figure 4
Healing of Haemorrhoissa, miniature from the *Codex Egberti*, Reichenau, 977–93. Trier, Stadtbibliothek, ms. 24, fol. 90v.

the hemorrhage is typologically combined with other biblical episodes and miracles related to fountains and/or fluidity. In the catacomb of Peter and Marcellinus, one of the Haemorrhoissas is placed to the right of Moses striking water from the rock, and the paralytic at the *Piscina Probatica* in Jerusalem (Figure 5; Baert 2002, 2005, 2007). The sixth-century encolpion from Adana also places the woman alongside the healing at the *Piscina Probatica* (Figure 6; Schiller 1971) as does a Carolingian ivory (Figure 7).[16] Another Haemorrhoissa in the catacomb of Peter and Marcellinus is combined with the Samaritan woman at the well (Figure 8). In the mosaic of Ravenna, the healing of the Haemorrhoissa precedes the Samaritan (Figure 9). On the sarcophagus in Santa Maria presso San Celso in Milan, the healing is combined with the apocryphal story of Saint Peter turning the prisoner's wall into a fountain (Figure 10; Knipp 1998: 124–5).

In other words, the iconography of the woman with the hemorrhage ties in with three fundamental categories in visual anthropology: the body, fluidity, and cloth. The role of textile in the Haemorrhoissa *Nachleben* is complex and multilayered. I therefore dare to speak of a "textilization" of the iconography of the Haemorrhoissa at four levels.

1. Textile and garment as the story-immanent motif.
2. Textile and cloth as pictorial metaphors and embodiment.
3. The role of textile in the material culture of the Haemorrhoissa.
4. The textile motif and the Haemorrhoissa in the debate on cult images.

Textile and Garment as the Story-Immanent Motif

Crucial in the text is of course the touching of the garment of Christ. On the Brescia Casket we see how the Haemorrhoissa is represented clearly taking the garment in her hands (Figure 1; Watson 1981). On the so-called "Tree Sarcophagus" of Arles (*c.* 360) we see the Haemorrhoissa huddled down, supporting herself on one knee at Christ's feet, with both hands grasping part of the hem (Figure 11).[17] The same is true for the Sarcophagus of Adelphia (*c.* 340) in Syracuse, Sicily (Figure 12), and the sarcophagus in the Leiden museum of antiquities (Figure 13).[18] In a Greek bible at the Walters Art Gallery it is, again, specifically the hem that the woman touches (Figure 14).[19]

The Gospels according to Luke (8:44) and Matthew (9:21) both, indeed, specify that the woman desired not simply to touch the garment of Christ, but more in particular the hem of the garment of Christ: *kraspedon* or *fimbria* (Fitzmyer 1985: 742–7). Touching the hem is a gesture one would expect from a hidden attempt to contact somebody. It implies a crawling, humble position. The hem psychologizes the Haemorrhoissa's touch as a clandestine act on the one hand, and as an act of submission on the other. Touching someone's hem is often a form of pleading with them.

Figure 5
Wall paintings with *Haemorrhoissa* in the catacomb of SS Peter and Marcellinus, third century. Rome.

Figure 6
Enkolpion of Adalana, sixth century. Assiût, treasury.

Besides the psychological impact, the hem has specific exegetical, material, and cultic meanings. Exegetes agree that the detail in Luke and Matthew is quoted from Mark 6:56, which says that wherever Christ went, people tried to touch the hem of his garment, and who did this, was healed (Cummings 1980: 51–2). The same idea appears later in Matthew 14:36 again: "And they besought him that they might touch but the hem of his garment. And as many

Figure 7
Healings and miracles of Christ, Court School of Charlemagne, ivory book cover, 211x124 cm, beginning of the ninth century. Oxford, Bodleian Library, Cod. Douce 176.

Figure 8
Wall paintings with Haemorrhoissa in the catacomb of SS Peter and Marcellinus, third century. Rome.

Figure 9
Christ and the Samaritan woman, mosaic, 520–6. Ravenna, S. Apollinare Nuovo.

as touched, were made whole." Patristic sources, moreover, saw a type of the healing hem of Christ in Zechariah 8:23: "In those days ten men of all languages and nations will take firm hold of one Jew by the hem of his robe, and say: Let us go with you, because we have heard that God is with you." The prophecy referred to the exaltation of Jerusalem, to the gentiles united in an eschatological pilgrimage. In the first Christian centuries, Zechariah 8:23 was already recognized as a messianic prophecy. Jerome (347–419) for example, interpreted the hem as the Church of the apostles (Jerome 1970: 822–4). "The miracle is understood as a symbolic action that is prophetic of the coming of the kingdom and the Parousia. The woman with the hemorrhage underwent her eschatological pilgrimage to God" (Cummings 1980: 53). The fact that the hemorrhage miracle is a sea crossing miracle (Struthers Malbon 1984: 363–77), was recognized as a "new exodus" and a "new Moses" theme. The story mirrors deliverance through water to newness of life, thereby suggesting a "baptismal as well as a Eucharistic, namely paschal, context" (Cummings 1980: 53).[20]

This would explain the popularity of the healing in paleo-Christian culture, a culture that was particularly sensitive to the approaching Heavenly Jerusalem. We can now understand that the eschatological concept of the hem as the very portal towards the new eternal life by way of water (baptism), and by way of communion (the blood of the Eucharist) led to the specifically *funerary* context, the catacombs,

and the many sarcophagi, where the Haemorrhoissa makes her appearance.

In short, to touch the hem is to be embraced by God, His kingdom, and the coming of Christ (Marshall 1989; Luke 6:19 and Acts 5:15, 19:12; Twelftree 1999: 133). According to Ambrose of Milan, this interpretation of the hem is particularly true for the gentiles who will become united in one church, as the church has blood wounds too, but desires to be cured.

But what do we know about the hem in Judeo-Christian material culture? The word *kraspedon* or *fimbria* refers to the fringes of the woven border (Haulotte 1966: 65). A fringeless hem would evoke the word *limbus*. The Book of Numbers (15:38) mentions the tassels or tuft: "Speak to the Israelites, and tell them to make fringes on the corners of their garments throughout their generations, and to put a blue cord on the fringe at each corner." Yahweh commanded Moses that the hem be bordered all around. These are the so-called Jewish Tsi-Tsit. The Jews would count the words of Yahweh from the numbers of the twisted white threads. It brought to mind the election and liberation of their people.

We should add also that the priests of Israel wore garments without seam—*seamless* is not to be confused with our issue, hemless (Edersheim 1951). Possession of the seamless garment of Christ, a symbol of the unity of the Church, mentioned in John 19:23 as the *chiton arraphos* or *tunica inconsultilis* for which the soldiers cast lots at the Crucifixion, is claimed both by the cathedral

(a)

(b)

Figure 10
(a) Healing of the Haemorrhoissa
and (b) Peter striking water from the
prisoner's wall, fourth century. Milan,
S. Maria presso S. Celso.

Figure 11
Christ and the Haemorrhoissa, detail from the so-called "Tree Sarcophagus," Gaul, *c.* 350–60. Arles, Musée Lapidaire d'Art Chrétien.

Figure 12
Christus and the Haemorrhoissa, detail from the sarcophagus of Adelphia and Syrakus, Rome, *c.* 340. Syracuse (Sicily), Museo Nazionale.

Figure 13
Christus and the Haemorrhoissa, detail from a sarcophagus, marble, h: 63.50 cm, w: 21.15 cm, Rome, 39. Leiden, Rijksmuseum van Oudheden, inv. no. Pb 35.

Figure 14
Christ and the Haemorrhoissa, wall painting in the catacomb of SS Peter and Marcellinus, third century. Rome

of Trier and by the parish church of Argenteuil (Aretz *et al.* 1996; Le Quéré 1997). This branch of tradition is not directly related to the Haemorrhoissa.

Let us return to the fringes. The tuft has a deep religious meaning in Jewish tradition, "A refinement that would not have concerned Jesus" (Robertson and Perschbacher 2004: 87). In Matthew 23:5 Jesus refers to the tassels of the scribes: "And all their works they do for to be seen of men. For they make their phylacteries broad and enlarge their fringes." The typical Semitic fringes or tassels are, however, not represented at the border of Christ's mantle, in the third and fourth centuries, when the image of the Haemorrhoissa emerged. The vestimentary characteristics are modeled upon the late-Antique and Hellenistic fashions then in vogue in Rome. Nevertheless the hem remained a meaningful part of textile, vibrating at the border between symbolic language and effective actions.

In any vestimentary culture, borders (such as hems, tassels, but also lace) fragment and defragment the sensitive symbolic markers of the body: the feet, the middle, the neck.[21] The lower hem of the garment is a border between one's own body and the body of the other. The hem is moreover the entrance and the exit of power par excellence, uploaded as it were, with the idea of the very transfer of *dynamij* in the story. This symbolizes the hem as a liminal zone where transfer and transpositions become potent between the "I" and the "other," between me and you.[22] Finally, the hem is also a thriving contact made with the earth. The hem therefore equates with the symbolic meaning of the foot, evoking sand, earth, filth even; in short, the expressions of humility and contact with the Lord (Dr Aigremont 1909: 21; Verhoeven 1956: 166f.). That humble contact at the Lord's service leads to the textualization of textile. Bordertexts are added precisely at these symbolic edges, transforming them into transit zones of plea, praise, liturgy, glory, tears, joys and submission, connecting textile with medieval *textacy* (see Gandelman 1986: 107–12 for hems as text-bearers).

Finally, concerning the cultic meaning of the hem, one particular ramification is of importance for our story. Plutarch (AD 46–120) writes of the emperor Sulla: "As she (Valeria Mesalla) passed on along behind Sulla, she rested her hand upon him, plucked off a bit of nap from his mantle, and then proceeded to her own place. When Sulla looked at her in astonishment, she said: 'It's nothing of importance, Dictator, but I too wish to partake a little in thy felicity'" (D'Angelo 1999: 98; Lalleman 1997: 358; Theissen 1983: 134). And she did, becoming his fourth wife. Secretly pulling a thread had to be excused. She fears being regarded as demonic. On the other hand he actually believes that the textile will transfer, and will share with her, the power of the body that it wore (Plutarch, in Theissen 1983: 134).

The similarities between the classical anecdote about Sulla and the Haemorrhoissa are striking. Also on the visual level, late-antique connections can be recognized in the emperor's restitution iconography. This iconography consists of an upright statesman with a kneeling, often female personification of the province whose gesture expresses a *clementia* or supplication.

David Knipp in his *Christus medicus in der Frühchristlichen Sarkophagskulptur* recognizes a Hellenistic interference, an *ekphrasis* in the paleo-Christian Haemorrhoissa. One of the five Haemorrhoissas in the catacomb of Peter and Marcellinus has caused confusion given the peculiar line between the hand of the woman and Christ (Figure 14). If this is a remnant of the Sulla story, which it really seems to be, then the mural demonstrates the highly hybrid character of first-century iconography. This is what Moshe

Barasch would call "energetic inversions" (Barasch 1987: 170). Energetic inversion is the power of attraction that a gesture—natural or conventional—is capable of exerting over the collective memory of humanity, even to the degree that it becomes a formal-artistic recipient for various emotions and their shifting interpretations across the history of art.

The borders of the cloth circumscribe the body in space. As contour of the body they write the body in their embodiment in time and space. The hem, moreover, emphasizes the body's gravitation; it links the body also to the earth, as feet do. This brings me to my second level of textilization.

Textile and Cloth as Pictorial Metaphors and Embodiment

It may be clear by now that our healing story is a healing in which textile forms a transmitter of power. But also the garments of the woman herself are particularly interesting. In this second level of textilization, I want to investigate how textile and cloth operate as pictorial metaphors and how they embody the meaning in the visual medium. It is worth noticing that the Haemorrhoissa is often represented enwrapped in clothing, as if she disappears in her own garments. This might point to her feelings of shame, her secret desire, and her being outcast. In fact in some interpretations this is also expressed in the small size of the woman, as in the bible of Otto III combined with the story of Jairus (Figure 15). The Haemorrhoissa is literally banished to the outer limits of the composition of the miniature: An outcast image in the image.

Figure 15
Healing of the Haemorrhoissa and
the daughter of Jairus, miniature out
of the *Gospel of Otto III*, Reichenau,
end of the tenth century. Munich,
Bayerischen Staatsbibliothek, Cod.
Lat. Cim 58, fol. 44.

In the Byzantine mosaics of Ravenna the Haemorrhoissa coincides with her own cloth (Figure 2). As an abstract form, a body that is more amorphous, a stain, she seems to embody the formlessness of flux itself (Demyttenaere 1990). In the fourteenth-century Kariye Djami in Constantinople, this embodiment of the fluid is emphasized by her contours, which mirror the profile of the ground (Figure 16; Teteriatnikov 1995: 171; Underwood 1966: 72–4).

She is crawling formless matter that seeks to become whole and healed in form. Some scholars point to the fact that this Haemorrhoissa forms the bridge between Jairus, the Old Covenant, and Christ, the New Covenant. This evokes again the authentic interpretation of the shift from Leviticus to a new era, with new rules, where the doomed are embraced.

In the *Didascalia Apostolorum*, a Syrian text from the third century, which is clearly addressed to a Jewish audience, the author tries to convince Jewish converted menstruants not to isolate themselves but to come to church. I quote the *Didascalia*: "Because also she with haemorrhages who touched the hem of Christ's garment was not isolated, but worthy to be forgiven of her sins" (26, 62, 5).[23] However, until well into the fourteenth century, contemporary with the mosaics in the Kariye Djami, the clergy would fulminate against menstruants

Figure 16
Christ and the Haemorrhoissa, mosaic, fourteenth century. Istanbul, Kariye Djami.

entering sacred space. Mattheus Blastares of Thessaloniki says in his canon laws (*Syntagma*, 1335): "The woman with a flow of blood did not even dare to touch the Lord short of the border of his outer garments."[24]

A strong opposition between female fertile and messianic sacred blood, between the female body and the sacred space is born and will meander through the Middle Ages. This, of course, opens the gendered blood research fields of, among others, Joan Branham and Caroline Walker Bynum. It is a field I will pass by.[25] Suffice it for the moment to say that all this seems to saddle the Haemorrhoissa with an almost unbearable tension:

her bloody presence in the sacred and liturgical sanguine spaces of catacombs and churches. One Haemorrhoissa in the catacomb of Peter and Marcellinus is positioned next to the *agape* with a female servant. In the Kariye Djami, she appears in the cupola above the Eucharistic altar. The paradox dissolves in the interpretation of the touched hem as the gesture that opens the heavenly Jerusalem, given the symbolism of fluids, of water and of the Eucharist on the one hand, and given the fact that the woman—also according to Ambrose (339–97)—is the embodiment of strong faith, of *pistis*. Ambrose says: "The likeness of the church is that woman who went behind and touched the hem of Thy garment, saying within herself, 'If I do but touch His garment I shall be whole'. So the Church confesses her wounds, but desires to be healed."[26] Whoever believes can rightly be called the hem of the church, concludes Ambrose.

The Role of Textile in the Material Culture of the Haemorrhoissa

In his catalog of Early Christian gems, Jeffrey Spier discusses a crystal in which the Haemorrhoissa is represented at Christ's feet (Figure 17). In his article *Medieval Byzantine Magical Amulets and Their Tradition*, the same author considers a silver intaglio (New York Metropolitan Museum, 5 cm)

in hematite with on the one side the Haemorrhoissa and on the other an orant, probably Mary (Figure 18; Spier 1993: 44, Fig. 6b, 1917, sixth–seventh century).[27] The epigraphy refers to the chapter in Mark (Frazer and Weitzman 1977: 440). Hematite, or bloodstone, has a red radiance that becomes black when it oxidizes (Meier 1977: 392–5). It was thought to be petrified blood, and from a medicinal point of view it was used to treat dangerous blood flows. The medieval mind recognized in the word etymologically *haima-tithenai*—"blood that stops."[28] The fourth-century *Orphica* says that hematite was formed when the blood of Chronos dripped down on the earth: *A leech come down from heaven* (Abel 1885 mp. 131n, verses 642 and further; Barb 1948: 67).

Another amulet preserved in a private collection combines the Haemorrhoissa with a snake's head (Figure 19; Spier 1993: 44).[29] The gem introduces the hysteria motif in connection with the hemorrhage (Veith 1965). The snake's head is the portrait of the uterus. The portrait will protect her, calm her down (Pointon 1986). In fact the uterus was considered an animal difficult to tame, that has the capacity to swell up, and to shrink. It was believed that the womb could develop tentacles that could take over the whole body, reaching the level of the throat (Leyerle 2001; van Loo 2007: 112–13).[30] In the lower material world of gems and tokens the *passus* in the Bible becomes the flipside of the intrinsically spasmodic character of the uterus and its phantasm of the

Figure 17
Amulet, crystal. New York, American Numismatic Society, inv. 307.

Gorgo or Medusa. More strongly put: the Haemorrhoissa *is* the uterus. She *is* hematite, animated by the intaglio that inscribes the healing power on the body.

Being born in a textual world, then entering the iconic world and the liturgical space, the Haemorrhoissa, finally, sank into the performative world of magic and body art. In that world, she actually returned to the liminal zones of tactility, the body, and the magical object worn on or under textile.

The Textile Motif and the Haemorrhoissa in the Debate on Cult Images

My final point treats the textile motif as impetus of the Haemorrhoissa's ramifications into the realm of the debate on cult images and *acheiropoietoi* or how textilization finally interweaves with the "iconization" of paleo-Christian culture.

An important source in that regard is Eusebius' *Ecclesiastical History*, book 7, chapter 18 (Eusebius of Caesarea 1955: 191–2; see also Bugge 1975: 127–39; Fabre 2007: 229–51; von Dobschütz 1899: 31, 197):

1. Since I have mentioned this city I do not think it proper to omit an account which is worthy of record for posterity. For they say that the woman with an issue of blood, who, as we learn from the sacred Gospel, received from our Saviour deliverance from her affliction, came from this place, and that her house is shown in the city, and that remarkable memorials

Figure 19
Amulet with hysteria motif. Private collection.

of the kindness of the Saviour to her remain there.

2. For there stands upon an elevated stone, by the gates of her house, a brazen image of a woman kneeling, with her hands stretched out, as if she were praying. Opposite this is another upright image of a man, made of the same material, clothed decently in a double cloak, and extending his hand toward the woman. At his feet, beside the statue itself, is a certain strange plant, which climbs up to the hem of the brazen cloak, and is a remedy for all kinds of diseases.

3. They say that this statue is an image of Jesus. It has remained to our day, so that we ourselves also saw it when we were staying in the city.

4. Nor is it strange that those of the Gentiles who, of old, were benefited by our Saviour, should have done such things, since we have learned also that the likenesses of his apostles Paul and Peter, and of Christ himself, are preserved in paintings, the ancients being accustomed, as it is likely, according to a habit of the Gentiles, to pay this kind of honor indiscriminately

to those regarded by them as deliverers.

The statue in Paneas is also mentioned in a homily of Basil the Great (329–79), bishop of Caesarea.[31] He specifies that Maximinus removed the statue, and that the statue was a speechless image animated by a miracle. And last but not least, Basil gives the Haemmorhoissa a name: Berenice. Berenice is also her name in the *Acti Pilati* (after 375). These Acts recount that when Christ went to the house of Pilate the twelve standards with the images of the emperor bowed to Him. In the same passage Pilate asks for witnesses for Christ's trial. One uninvited woman, Berenice, shouted her story from a distance in the crowd. The Jews, however, refuse her oral and female testimony. Berenice would refer to the carriers of the banners, *pehre-nike*. The twelve standards refer to her twelve years of illness, and also here she is hidden in the crowd. What is also striking is the introduction of miraculous images (Kuryluk 1991: 119–20). The image cult of the ancients begins to shift towards, to bow to the new image cult of Christianity.

Philostorgius (368–430)[32] says in his *Historia Ecclesiasticae, VII,* 3 that the people of Paneas honored a statue of Jesus that was placed near a fountain on the market place by the hemorrhaging woman herself (Wolf 2002: 287). He adds that this place was very suitable for a woman that suffered from her menses. It was, however, destroyed by Julian the Apostate (361–3) (von Dobschütz 1899: 94). Philostorgius mentions that a head with a golden glow was preserved and venerated in the city. A sixth-century pilgrim reports on a golden-bronze image, but after that the object disappears from the testimonies.

The story of the statue in Paneas marks the transition in iconic systems. Eusebius and Basil hold different opinions. Eusebius is wary of the veneration of images by Christians. Basil, on the contrary, sees the image as a moving testimony to a miracle. And the later important defender of image cult and veneration, John Damascene (676–749), refers to the statue of Paneas as an argument for iconophilia.

Further legends embroidered based on the *Acti Pilati* or *Acts of Pontius Pilatus*.[33] From the fifth to the eighth century the *Cura*

Sanitatis Tiberii took form, probably arising in Northern Italy (perhaps Tuscany), closing the circle of Haemorrhoissa—Berenice—Veronica.[34] The legend relates how the Roman Emperor Tiberius, seriously ill, had heard of Christ. He decided to send for this healer/Savior. Volusianus, a high imperial official, was sent out, and reached Jerusalem after a year and three months. Pilate and the Jews were terrified by the summons, as Christ had in the meantime been crucified. The people blame Pilate and he is imprisoned. Then Volusianus hears of a woman from Tyre who was cured of an issue of blood by Christ, and had an image of him made. He sends for this woman, examines the image, and orders that everybody implicated in Jesus' death be killed. He then takes Pilate, the healed woman, and her image, with him on his return journey, but banishes Pilate from Rome. Upon looking at the image of Christ, Tiberius is cured. An eighth-century gloss identifies this anonymous woman as Veronica.[35] "The legend combines the *Acts of Pilate* with the legend of Paneas. The Paneas legend speaks of a statue, while the *Cura Sanitatis Tiberii* suggests an icon. It is possible that the *Cura Sanitatis Tiberii* adapts the nature of the portrait in order to make it transportable" (Kusters 2009). Furthermore, from the sixth century onwards the Syrian Legend of Abgar was spread, recounting how the king of Edessa was, in a similar fashion, cured of leprosy thanks to a cloth bearing the imprint of Christ's features: the *mandylion* (Wolf *et al.* 2004). All these tributaries feed into the Roman Legend of St Veronica. In the twelfth century the first testimonies occur of a *sudarium*.[36] Gervase of Tilbury mentions an image *secundum carnem*, apparently a bust painted by Veronica and preserved in San Pietro (von Dobschütz 1899: 292). Gerald of Wales says that he saw in Rome a veil (*peplum*) with the imprint of Christ's face (*impressit*): *vera icona, id est, imago vera*. According to the same author, nobody saw the actual image since a veil (*vela*) hung before it (Cambrensis 1873: 274ff.).

To return to the statue of Paneas, most probably the statue was a late antique group, adopted by the Christians with a new content. Thomas Mathews, in his *The Clash of Gods*, shows a coin with Hadrian and Judea in analogous postures: the kneeling, the outstretched arms (Figure 20; Mathews 1993: 62, Fig. 42). Because of the mention of a tonic plant in Eusebius' account, some authors presume that the statue originally represented Asclepius, perhaps with his daughter Panakeia who, according to the story, searched and found a healing plant (Eisler 1930). This would mark a second transition: the transition from Greek herbal therapy into the miraculous healing by Christ.

In sixth-century Greece and Mediterranean Italy, incantations mention the herb *berenikij* that has many healing powers. In the same group of charms *berenikij* is related to blood demons. "So thou too stop the blood of the servant of God, congeal disease, as that one and as a Stone, may it be annulled. I exorcise thee by the Faith of Beraioonikij blood, that you may not drip further; let us stay good, let us stay in fear; amen.

Figure 20
Coin with Hadrian restoring Judea.
Naples, Museo Nazionale.

Jesus Christ conquers" (Barb 1948: 38–42; Ebermann 1903).[37]

Botanists have traced the plant back to Persia. It is indeed a so-called border plant that prospers in dry, rocky ground, under the name *veronica chamaedrys* or *spicata*. In English cottage gardens it is still to be found under the name *speedwell*.[38] Indeed the herb was said to have a tonic effect on blood. Whatever it is, the officinal herb described by Eusebius, climbing up from the socle, over the feet to the hem of Christ, seems to mirror the Haemorrhoissa herself: creeping, crawling, clinging to the earth, but moving towards that desired powerful hem. And is it not remarkable indeed, that the Indo-European linguistic root of the hem—*bherem*—means, besides to seam of cloths, common to the bank, the edge of the love bed, and thirdly, a border plant.[39]

The Anthropological Space: Threads and Roots

At this point, we have reached deep archetypical contrasts between living herbs and inert dead metal, between flux, such as the pouring hot metal and the fountain on the market, and the dry counter-flux in the statue and her healing. These metamorphoses in nature are the fundamental principles of pre-medicine cosmology. "Living in close contact with plants and animals, the pre-modern human being had developed a three dimensional system (if we add rocks and minerals, we might say a four-faceted system) in which creation (or, if you will, the existence—everything organic, animate, living, and sentient) lived together in the interaction of an ongoing exchange of perceptions, information, and fluid or ethereal substances" (Camporesi 1995: 88). We have seen that in this context, the hem is a universal vibrating transit symbol that, besides the nature of "water" and "plant," remains connected with the very border of the textile. From the narrative and the iconic space, my reading of the hem is as the symbol of demarcation and discontinuity. This discontinuity was part of the healing miracle: a flow was stemmed. The syntax of the interruption was translated into the medium and the symbol of the hem. However, the hem as "discontinuity" is not solely a rupture, but also a locus of exchange: the locus of the transit of power (*dynamij*) from one person to another.

In this last phase, we touch a methodology that goes beyond iconographical and literary sources, but departs from ethnographical knowledge to reach the deepest archetypes.

Three angles will be adopted: the semantics of the hem and its link to generative capacities, the cosmology of gnosis and the cosmogonic womb, and the anthropology of the textile arts as a lead to an increasing iconization of a uterus phantasm. From these three angles I will approach the point of intersection between the Haemorrhoissa and Veronica, to whose figure the woman with an issue of blood was assimilated during the Middle Ages. For is the fundamental question not: how could touching the hem of Christ's garment lead to the woman who carried the imprint of Christ's face? An answer is possibly to be found in the anthropological interweaving that displays itself between the Haemorrhoissa, the touch of the hem, the uterine blood, and the emergence of visuality.

In her *Saum und Zeit*, Ellen Harlizius-Klück builds from a metonym: the interpretation of the body of the Virgin in light of her theological significance as origin of the era of the New Covenant and the location in which the Incarnation took place.[40] These semantics link the hem as the visual contour of the body, and the concepts of time and space. Within this analogy, the hem acquires the force of a *pars pro toto*. The contour of the body itself forms a temporal and spatial "perigraphy." The power of the concept of the

hem is also apparent from the many linguistic prototypes and offshoots. Harlizius-Klück's research grew into a lexicon, with the author tracing, amongst other things, the Hebrew denotations that connect the hem to the bosom and, more importantly, the womb.

Heik is the Hebrew for both hem and bosom. The bosom is understood as the marked opening in the cloak, just above the belt (Is. 40:11, Ps. 74:11) (Harlizius-Klück 2005: 94–5). *Kanap* means wing, cloak hem, hem, edge, but also lap and womb. It is related to the cloth that covers the private parts (Deut. 23:1; Deut. 27:20). In Ruth 3:9 the taking of the father's wife is compared to the taking of the father's blanket or cloth. It is, indeed, this root that led to the typical Hebrew cloak—in fact a draped blanket—with the tassels already discussed: the Tsi-Tsit. As has already been said, this cloak/cloth brings to mind Yahweh's commandments and the tassels relate to the laws of the Old Covenant (Num. 15:38; Deut. 22:12) (Harlizius-Klück 2005: 105). *Sul* is the hem of Yahweh's royal garb. Isaiah 6:1 says that the hem filled the Temple. The word also refers to the edge and circumference of the pitcher, in its turn an image of a woman's lap. The pitcher is a universal symbol for the preservation of the mystery and the metamorphosis that will take place (Lurker 1991: 232–3; Neumann 1956, 51ff., 123–46). The *vas* is an image of the womb in both form and usage. The Song of Songs sings "Your navel is a rounded goblet" (7:2) (Baert 2006: 35–62). In combination with the verb *galahn*, *Sul* is to uncover (Jer. 13:22,

25–6). To uncover the hem implies revealing a vulnerable part of the body (Nahum 3:5) (Baert 2006: 187). This explains why the hem is also related to erotic feelings and feelings of shame. In the Legend of the True Cross and the Queen of Sheba, we still encounter traces of this. In a rhymed German version by Heinrich von Freiburg, dating from 1275, we read:

> *Da beugte sich die Herrscherin,*
> *kniefällig betend fiel sie hin,*
> *und Lüftete den Saum der*
>   *Kleider*
> *mit blossem Fuss, den sie*
>   *enthüllt.*
> *Darauf sprach sie, vom Geist*
>   *erfüllt,*
> *demütlich das haupt gesenkt,*
> *und zu sich selbst dazs Wort*
>   *gelenkt:*
> *Das Zeichen des Gerichtes ward*
> *vor meinem Blicken offenbart.*
>   (Beyer 1987: 242)

Then the queen bowed,
on her knees she fell
and lifted the hem of her dress
and a naked foot she unveiled.
Then she said, filled by the
  Ghost,
Her head so humbly,
and she spoke to herself:
the sign of the Judgment
has appeared before my gaze.

The "unveiling" of the foot by the hem is immediately followed by the "fulfilment" of insight. In a subtle manner, the author links the mysterious disclosure of the foot with eroticization, humility, and revelation.

The Hebrew roots of the concept of the hem provide a number of links connecting the covering of the most intimate, to the womb and by extension to sacred space. For Harlizius-Klück this indicates that in its earliest form the hem was not a circumscription in the meaning of a discontinuity, but in the sense of the world's "circumference," and thus related to the continuous and to eternity. In its authentic semantics, according to this author, the hem therefore has a much more spatial than temporal texture. In Ellen Harlizius-Klück's words:

> *Die These ist, dass die*
> *Bedeutungslage des Saumes*
> *als Schnittkante, Rand,*
> *Marginalie und Diskontinuität*
> *und die der Zeit als konstante,*
> *kontinuierliche und ordnende*
> *Kategorie ehemals umgekehrt*
> *war: Der Saum stand für den*
> *ordnungsgebenden Anfang und*
> *die Kontinuität der Tradition;*
> *die Zeit wurde wahrgenommen*
> *in der Form eines Einschnitts,*
> *der Wunden im Leben erzeugte,*
> *die es durch bindende Gewebe*
> *zu heilen galt. Es stellte sich*
> *im Laufe der Untersuchung ein*
> *Motiv heraus, und dem sich*
> *die Umkehrung am besten*
> *darstellen lässt, das aber als*
> *begriffliches Instrument in den*
> *konsultierten Wörterbüchern*
> *fehlt oder als Randerscheinung*
> *angesehen wird: der Saum als*
> *genealogischer Ort.* (Harlizius-
> Klück 2005: 5–6; emphasis
> added)

The hem refers to the place of (pro) creation itself: the womb, the space of the holy of holies. As a locus, the hem frames a time that is felt as a wound, an opening, and a cut in the fabric.[41]

From the perspective of Ellen Harlizius-Klück's *Saum und Zeit* we can therefore derive a new hermeneutics of the hemorrhaging woman.[42] The Haemorrhoissa touches the hem: the very matrix of time and space. The wholeness of the healed Haemorrhoissa is mirrored in the hem as symbol and image for completeness, sacred space, restored genealogy, and procreation. Her touch is a metaphor for the old verbal images of the womb, the pitcher and the Temple. Her touch is located in the zone of the veiled, and therefore partakes of shame throughout the unveiling. One could say that in this hermeneutic, the hem is the second form of the Haemorrhoissa herself—vase, uterus, shame—but also of Christ—the Temple. It is precisely in the hem as metonym that these connotations flow into a single *dynamij* that jumps back and forth between Christ and the Haemorrhoissa. According to this hermeneutic, the hem loses its connotation of separating boundary, of "discontinuity" as rupture, and tips into a different type of dynamic, namely one of both sides.[43]

In the second approach—the gnosis—we will see that the hem, and by extension the female lap as *ordnungsgebenden Anfang* (Ellen Harlizius-Klück) entered and spread through the cosmological outline in other contexts as well.[44] In the second-century Gnostic literature the biblical Haemorrhoissa appears under the name Prunice, regarded as the feminine principle of the universe and identified with Sophia, "wisdom," and Mother Achamoth, the "second Sophia," the more negative side of true wisdom (Kuryluk 1991: 107–11).[45] Prunice's flow was seen in analogy and contrast to the flow of Christ's blood; the divine emanations as seed and Prunice's flow as procreation. In Valentinianus' Gnosticism, known through Irenaeus' treatises (late second century) and Epiphanus (*c.* 315–402), Prunice is "bearer," the twelve years of her bleeding interpreted as twelve emanations or children.[46] The Prunice-mythology reflects the Gnostic idea of a variety of worlds, the so-called Aeons, divine emanations or thoughts, the best-known of which is the Aeon Sophia. For the Ophites she is the emanation of the first woman, mother of all living things and bride of God the Father and the Son: she is responsible for the birth of Christ.[47] "In the Gnostic cosmologies the twelve years of hemorrhage are transformed into the apocalyptic twelve-year passion flood that would—had Christ not come—have destroyed the earth. But when she touched the hem— symbol of the moon's last quarter— the flood was stopped and the world was saved from destruction" (Kusters 2009).[48]

In the suppressed dualistic cosmology of the gnosis, the Haemorrhoissa is anchored in the archaic anthropology of blood: the opposition between menstrual flow and internal blood, between *menses* and seed, between the feminine and the masculine entity, between unclean blood and pure blood, between negative pole and positive pole (Camporesi 1995: *passim*).[49] "In both its positive and negative aspects, it represents carnal existence. Menstruation opens the way for the possibility of life in this world and is an apt symbol for the messy flux or mortal flow of life" (Buckley and Gottlieb 1988: 77). In short, at least in dualisms such as gnosis, the Haemorrhoissa became connected to very old principles of potentiality, restoration, and the unique procreative capacity—paradoxical and anxious—of the bleeding womb. One need only think that the womb is—besides the nose—the only opening that sheds its own blood (benignly or malignly) with any frequency.

Traces of the matrixial primal principle were expounded by the early Church Fathers. Irenaeus (98–177) says that the descendants of Cain regarded the *Hystera* (the womb) as their creator (Barb 1953: 210). Hyppolitus of Rome (early third century) also comments on the matrixial cosmogony of the gnosis (Culiano 1987). The womb derives from the first congress of the two principles with the navel in the middle: the microcosmic center, the *omphalos*. The uterus is characterized by black (sometimes "dirty") underground flows. It is the *prima* or *diva matrix*, still inert and in a certain sense genderless. The dark water, however, continually desires to be united with the light of the skies. This struggle gives rise to all others, and to gendering. Before the heavens and the earth were created, says Genesis, God's spirit moved over the waters.

Ideas about the *diva matrix* as "creatress" were still known in medieval Europe. Paracelsus calls dark water—preceding the Creation and uniting itself to the divine spirit—the matrix. In this womb and no other heaven and earth are created, says the sixteenth-century scholar (Paracelsus 1922–33: 177f.). The ancient *diva matrix* symbol— always with a curve that ends in two bulges, whether or not snake-like or spiral in shape, and often similar to an octopus—were disseminated on gems that were known in Europe into the seventeenth century. We know from N. C. F. Peiresc's correspondence that Rubens owned such a gem (Barb 1953: 194). The art dealers had recommended a Gnostic amulet to his attention: an *intaglio d'amethista con la vulva deificata et revestita della le di farfalla*. Around the uterus was the inscription THEAZ MHTP (*diva matrix*). In his reply, Rubens expresses his excitement at the tip and in the margin he even sketches the gem (Figure 21).[50]

As we have already seen, the Haemorrhoissa gems also fit into this fickle world of the evocation, conjuration, and glorification of the awful uterus. This shows once again that the biblical bleeding woman tied in with patterns of feeling that preceded the composition of the biblical text or could be explicated in alternative genres of the time, such as the gnosis. The biblical narrative guise initially thematized in the synoptic Gospels as a miracle of healing, was sucked into the hidden uterine models from the world of Gnostic cosmology, on the one side, and the material cultural world of magic on the other, where it could swell into a new archetype. This process of recognition and appropriation from the Bible indicate a boundless energy that remained "beneath" the narrative and iconic spaces of the Christian spectrum.

This brings me to the third angle of approach: the anthropology of the textile arts as a lead to the increasing iconization of a uterus phantasm.

One can historically identify a desire to iconically grasp the unfathomable regenerative powers of the matrix or uterus, the impalpable "blackness" of her underground but nevertheless life-bearing flows. This led to both aniconic forms and motifs evoking a feel of this unfathomable potential as well as phantasmatic images increasingly entering an iconic register.

The case of Berber weaving offers an approach to this first issue, and offers one a view on the primal expressions of what would culminate in an iconization of the uterus phantasm (Vandenbroeck 2000: *passim)*. The inner and material culture of Berber women has remained relatively uncontaminated by masculine thought patterns. Their ceramics and textiles furthermore show a symbolic and even proto-symbolic expression that, with minimal "pollution," goes back to the Neolithic. For Berber women, the act of weaving is a cosmogony and a procreation (Figure 22). Weaving itself mirrors the potentiality and power of creation from fluid, dark nothingness. The warp yarns are

Figure 21
Sketch of the gem in possession of Peter-Paul Rubens. From Barb (1953: 193–238, nr. 25d).

the *axes mundi*, and the cutting of them is compared to birth itself, with the cutting of the umbilical cord. The "hems" or fringes that arise are reminders of the separation between mother and child. The colors used are primarily black and red. In this cosmo-analogy of weaving a particular motif is central: the ever-recurring central iconic hollow in the cloth, often a stain-like black spot, a place to which the weave tends, something that pulses in the fabric. Paul Vandenbroeck calls this the "nameless motif." Berber women explain the recurrent black spot as the shuddering spasm of the uterus itself (Vandenbroeck 2000: 114).

In fact this "nameless motif" is a prefiguring of the later uterine icons, as seen in the gems. The uterine icon as such develops from the amorphous, the fluid, and the stain-like to the figurative. This path goes from a positive fascination to a more fearful one. Octopus-like creatures, for instance, or balls with pins, tentacles like hedgehogs, and snakes' heads are often found as figures of the womb, corresponding to the swelling and shrinking capacity that the womb was thought to have (see also van Loo 2007: 113f.). In the last phase of the growing figuration, it was even given a face: that of Medusa.

The representation of the matrix as a face has its roots in the pre-classical pantheons and myths.[51] The Babylonian mother-goddess Nintu lives in the mountain[52] and embodies fertility and regeneration. Her face is represented with the abstract sign of the female sex: an arch ending at the bottom with a curl similar to the Greek letter Omega (Figure 23; Barb 1953: 199). The same shape—schematized from the female anatomy—also appears for the Egyptian mother-goddess Hathor (Figure 24). As a hieroglyph the sign means "protection" (Murray 1911).[53] There are archaeological finds that suggest the possibility of Medusa being assimilated to Hathor (Nintu) (Pettazzoni 1921). An Etruscan terracotta from Campania shows that the apotropaic face is placed within the omega-scheme with a shell as background (Figure 25). In brief, a route can be traced from black formlessness to schema to zoomorphic symbol (octopus, hedgehog), and from the animal to the anthropomorphic.

The Medusa with snakes for hair, but also the Haemorrhoissa

Figure 22
Berber textile, date unknown.
Private collection.

who joins Medusa on the medieval gems, forms an extreme stage of the increasing iconization of the uterus phantasm. Thus the apparently immeasurable distance between flux and face, between textile and Medusa, becomes bridgeable. An impressive continuity is revealed between the fluid blackness and the apotropaic face of the *diva matrix*. This perhaps identifies the point of intersection between the woman who touched the hem of Jesus' garment and the woman who would carry the face of Christ.

Figure 23
Babylonian relief showing Nintu. From Barb (1953: 193–238, no. 27c).

Figure 24
Symbols for Nintu and Hathor. From Barb (1953: 193–238, no. 28c).

Figure 25
Terracotta showing "Medusa,"
Etruscan, *c.* 500 BC, Campania. From
Barb (1953: 193–238, no. 33a).

At a micro-level we have already seen how in the third century the Haemorrhoissa was given a name—Berenice—and that from this naming, traces were drawn between a woman cured miraculously and a specific witness during the trial of Pilate. In the Acts of Pilate this woman is put forward as the source of the iconic witness to Christ. At this level, the Haemorrhoissa can be read as the starting point for the defense of a Christian visual culture. At a macro-level we are dealing with the connection between the uterine taboo and the unveiling of Christ's face, that implies that Christianity—the Medusa as well as the *vera icon*, the true face of Christ, warding off evil—has never lost its fascination for the feminine as *apotropaion* (Gaignebet 1976).

But that also entails the working through of a hidden history in which menstruation, textiles, and the frontal face are knit together. Or to put it another way: the source of the iconic witness in some manner lies in female fluidity.

Final Words

The flow cured by Christ, the interruption in the stream, created the space for the image as living memorial (Figure 26). It again brings to mind the cut in the cloth, the sacred place encircled by the hem. Between the quarters of a lunar cycle, an image comes forth, potential, flowing, triumphantly displayed and borne on textile: the *vera icon*. How can we understand the origins of the image in the flow of blood otherwise than as a *hysto*logy?

And should we in consequence not see this face that manifests in textile as the image of the figurating self: *hysteria*? Between the Haemorrhoissa and Veronica unfolds the mysterious ellipse that ties together fabric and making flesh.

"For it was you who formed my inward parts; you knit me together in my mother's womb [...] when I was being made in secret, intricately woven in the depths of the earth" (Psalm 139:13, 15). "The Word became flesh" (John 1:14).

The memory of the face of God could come about through the touching of a hem.

Acknowledgments

This article is part of the research project "The Haemorrhaging Woman (Mark 5:24–34parr). An iconological research into the meaning of the bleeding woman in medieval art (fourth–fifteenth century). Also a contribution to the blood and touching taboo before the era of modernity" funded by the Research Fund of the Katholieke Universiteit Leuven (2008–12). I am grateful to my PhD students, Liesbet Kusters and Emma Sidgwick, and to Gerhard Wolf for the inspiring conversations.

Figure 26
Saint Veronica with the Sudarium, Master of Saint Veronica, *c.* 1420. London, National Gallery.

Notes

1. The Haemorrhoissa is treated in the following exegetical studies: Bovon (1991), D'Angelo (1999), Fonrobert (1997), Taylor Gench (2004), Haber (2003), Horsley (2001), Lemarquand (2004), Levine (1996), Marcus (2000), Melzer-Keller (1997), Metternich (2000), Oppel (1995), Plaskow (1993), Schüssler Fiorenza (1994), Selvidge (1984, 1990), Söding (1985), Struthers Malbon (1992), Trummer (1991), Wilkinson (1998).

2. English quotation from the New Revised Standard Version (NRSV). The Vulgate reads:

   et abiit cum illo et sequebatur eum turba multa et conprimebant illum et mulier quae erat in profluvio sanguinis annis duodecim et fuerat multa perpessa a conpluribus medicis et erogaverat omnia sua nec quicquam profecerat sed magis deterius habebat cum audisset de Iesu venit in turba retro et tetigit vestimentum eius dicebat enim quia si vel vestimentum eius tetigero salva ero et confestim siccatus est fons sanguinis eius et sensit corpore quod sanata esset a plaga et statim Iesus cognoscens in semet ipso virtutem quae exierat de eo conversus ad turbam aiebat quis tetigit vestimenta mea et dicebant ei discipuli sui vides turbam conprimentem te et dicis quis me tetigit et circumspiciebat videre eam quae hoc fecerat mulier autem timens et tremens sciens quod factum esset in se venit et procidit ante eum et dixit ei omnem veritatem ille autem dixit ei filia fides tua te salvam fecit vade in pace et esto sana a plaga tua.

3. Charlotte Fonrobert calls these tendencies the *Jesus-was-a-feminist syndrome* of the nineties. Fonrobert (1997: 124) refers to Selvidge (1984), "androcentric levitical writers," Barta (1991: 31), and Swidler (1971: 181).

4. See also Neusner (1973: 65), Parish Sanders (1998: 214–30, 1990: 258–71), Fredriksen (1995: 18–25, 42–47, 22–23), and Gilders (2004). According to D'Angelo (1999) there is no evidence that the touch of a menstruating woman's hand or brushing up against her in a crowd would have been considered polluting in the first century.

5. See Metternich (2000: 78–131) and Fonrobert (1997: 130): concerning *zabâ (zvaha)* and Leviticus: "The menstruant woman does transfer impurity by being touched (Lev. 15.19). However, Leviticus does not mention that she communicates impurity by touching. Milgrom comments that this 'can only mean that in fact her hands do not transmit impurity [...].'" The Mishna is more severe: "He who touches a *zav*, or he whom a *zav* touches, transfers a status of impurity to food, drink and vessels that (can be purified by immersion)" (Fonrobert 1997: 131). But Leviticus does not say anything about the

zabâ who touches another body. Only in the later Mishnah (sixth–seventh century) is the touching *zabâ* explicitly a transmitter of pollution and enjoined to live in isolation in the Beraita de Nidda (Biale 2007; Branham 1999; Buckley and Gottlieb 1988; Cohen 1991: 278; Fonrobert 2000; Lemay 1990; McCracken 2003; Walker Bynum 2007; Whitekettle 1996; Wood 1981).

6. Rhoads (1994) is seen by some authors as belonging to the same category, but in this case the Greek woman throws herself at Christ's feet to demand the healing of her daughter (Mark 7:24–30).

7. I refer to our research project *Mary Magdalene and the Touching of Jesus. An Intra- and Interdisciplinary Investigation of the Interpretation of John 20, 17,* funded by the Foundation for Scientific Research—Flanders, in the department of Art History and the faculty of Theology at the Catholic University of Leuven (2004–8). See the publications on touch, the senses, and *Noli me tangere*: Bieringer (2006, 2008a, b) and Baert (2008: 255–308, 2009b, c).

8. See also Grob (1967: 85–6) and De Bruyne (1943: 166–74).

9. See also Baert (2005).

10. This contrasts with Fonrobert (1997: 127, n. 17), who says that "[...] healing by touch is also a common element in Hellenistic healing stories."

11. So-called *mana*: a place in his theory of the relationships between representatives of the spiritual world and mankind (Hull 1974: 87, 108).

12. Examples taken from Lalleman (1997: 355–6).

13. Ginzburg (1998: 101): "Ces images, concentrées sur le punctum, sur l'instant décisif."

14. On these confusions, see Knipp (1998) and De Bruyne (1943: 166–74). Due to the sometimes ambiguous postures, some of the secondary literature confuses the Haemorrhoissa with the *Noli me tangere* motif, for instance in the catacomb murals, the Brescia Casket and the so-called Brivio Casket (this last in Noga-Banai 2008: 38–61, Fig. 3). In these three cases no disciples are shown to be present. Christ's gesture to the kneeling woman reaching towards his hem can be read either as a tactile gesture (compare the *Chairete*) or as a gesture of rejection—*Noli me tangere*. Christ's turned torso can also create confusion with the turning away of the *Noli me tangere*. We are convinced that an independent iconographic motif for the *Noli me tangere* arose only from Carolingian times onwards.

15. Trier, Stadtbibliothek, codex 24, fol. 91 (Schiel 1960; Ronig 2005: 78ff.).

16. *Healings and miracles of Christ*, Court School of Charlemagne, early ninth century, ivory book cover, 211x124 cm, Oxford, Bodleian Library, Cod. Douce 176.

17. *Christ and the Haemorrhoissa*, detail from the so-called "Tree Sarcophagus," Gaul, *c.* 350–60. Arles, Musée Lapidaire d'Art Chrétien.

18. *Christ and the Haemorrhoissa*, detail from the Sarcophagus of Adelphia, Rome, *c.* 340. Syracuse (Sicily), Museo Nazionale; *Christ and the Haemorrhoissa*, detail from a marble sarcophagus, Rome, 390, h.: 63.50 cm, w.: 21.15 cm. Leiden, Rijksmuseum van Oudheden, inv. no. Pb 35.

19. Baltimore, Walters Art Gallery, Ms. Gr. 539 (Boeckler 1961: 9, Fig. 11).

20. So the story in Mark was to be interpreted also in a catechetical sense.

21. On this see Vandenbroeck (2009) and Warwick and Cavallaro (2001).

22. In the ancient Greek, *dynamij* means "potentiality" (Agamben 1999; Sidgwick 2009).

23. Translated from Connely (1929: 129, 254), Fonrobert (1997: 122), and Vööbus (1979).

24. Alphabetical Collection, A. 16, quoted in Viscuso (1991: 401).

25. The relationship between sacred space and the Eucharist goes beyond our present theme. See Walker Bynum (1992: 99, 2007).

26. Ambrosius, *Expositio Evangelii secundam Lucam* VI, 56–9, CCL, XIV, IV, 193–5; full text in Knipp (1998: 112–13).

27. Also shown by Kötzsche (1979: 440) and Nauerth (1983: 560, cat. no. 165). Spier (1993: 44, n. 11) also refers to the Benaki Museum in Athens, which preserves a Middle Byzantine green chalcedony gem showing

the Haemorrhoissa and the Crucifixion without inscription. Hematite is an iron ore that is not especially rare. This type of stone is characterized by a red core, which when worked (ground, polished) becomes black or silver.

28. Meier (1977: 394): "*Hematites (...) dicitur ab hema, quod est sanguis, et tithein, quod est sistere, quasi sistens sanguinem,*" following Petrus Berchorius (*c.* 1290–362), *Reductorium morale* XI, p. 440a. In the same passage, Berchorius treats at some length of blood-flows due to "*luxuria,*" "*carnalis voluptas, mundana prosperitas, fluxusque cujuscunque iniquitatis,*" referring to the passage in Mark. "*Figura de haemorrhoissa, quae ad tactum vestimenti Christi a fluxu sanguinis est sanata. Vestimentum Christi est abstinentia, quae re vera sanat ab istis fluxibus animam peccatricem.*" The relationship between this disease and sin needs further investigation.

29. Spier (1993: 44): "The bronze token with the haemorhoïssa suggests that it had to help women in some way."

30. The throat is a "shaft," a gorgo (see below), a channel that in early-medieval medical and magical thought was considered a mirror of the uterus.

31. Photius, ninth century (Fabre 2007: 237, n. 19).

32. Philostorgios *Historia Ecclesiasticae, VII,* 3. See von Dobschütz (1899: 199–200,

n. 1, and the original text on 257–9). See also Bidez (1913: 78–81).

33. See for these legends: Kusters (2009).

34. See for this legend von Dobschütz (1899: 209–212; and for the original text pp. 157–203); Kuryluk (1991: 120).

35. "*Veronica quae basilla [...] dicitur*" (von Dobschütz 1899: 210, n. 1).

36. There is abundant literature on the Roman *vera icon.* I have used here: Dragon (1991), Kessler (1998), Koerner (1993: 80 *passim*), Morello and Wolf (2000), Wilson (1991: 112–13, Fig. 10, ill. 14, 21, and 111), and Wolf (1998, 2002: 50–2).

37. The daughter of the Canaanite woman (Mt 15:22) is also sometimes called Berenice. In the Pericopes of Henry III (1039–43), Bremen, Staatsbibliothek M. 21, fol. 28v, the woman with an issue of blood is shown at the bottom, and in the upper register the episode of the Canaanite woman. In this episode, again, Christ is approached from behind.

38. Mentioned in Pradel (1907: 111).

39. Available at: http://www. indo-european.nl/. Another association is the intimacy of the edge of the bed, on the one side, and the shore of an island, on the other. There is perhaps, then, an association with the shore of the Sea of Galilee, which in the exegetical literature is also seen as a boundary and transit zone in Mark's redaction. See Struthers Malbon (1984).

40. With thanks to Prof. Dr Silke Tammen and Prof. Dr Felix Thürlemann, who drew my attention to this research (Harlizius-Klück 2005).

41. There is no room here to develop further the association between opening, tear and the temporal epiphany in the iconography of the Annunciation. In his *Le détail*, Daniel Arasse fixes our attention upon the Annunciation by Filippo Lippi (*c.* 1450) in London's National Gallery, in which at the height of the navel a buttonhole was painted with no button. The tiny opening is barely visible to the naked eye. As Arasse studies several almost-invisible details in his book that—as the author develops—should be considered aspects of the enigmatic and intimate interconnections between the artist and his work of art, this particular lost button could enclose a symbolic meaning. The little tear, opening precisely on a single horizontal line with the dove, the power by which the Word was made flesh, is, for whoever sees it, a subtle suggestion of the navel as the paradox of a closed opening. It is virginal, as the iconography requires, but also sensual and erotic like the work of art. There are authors who would identify the model for this woman as Lucrezia Buti, a lady at the court of Prato, where the *cintula*, Our Lady's girdle, was preserved. The girdle with a single large knot around the navel symbolizes the binding and unbinding of fertility. We know that Filippo Lippi had an unbridled passion for this woman. A buried detail is the visible invisibility of Lippi's pictoriality itself. "*Le désir du peintre dans la peinture même*" (Arasse 1996: 340). On the navel, see Bronfen (1998: 26–30) and Baert (2009d).

42. On biblical hermeneutics and "lire au-delà du verset" (Lévinas), see Ouaknin (1994: 37, 136, 283).

43. Stated in yet a different sense: the hem as locus of this particular dynamic, as a liminal zone where transfer between the "I" and the "Other" are able to originate, is a locus of a certain transformational potentiality. This conception of the hem could be said to correlate to, or be a concrete and culture-specific expression of, what was psychoanalytically (or, even better: from the perspective of a correction on classical, orthodox Lacanian psychoanalysis) identified as "matrixial borderspace," a transsubjective psychic sphere, that can be grasped according to the model of the maternal womb (matrix) and its co-emergence-in-differentiation (fusion nor separation) of an I and an uncognized non-I, and their "borderlinking" as a nonfusional transmission, connectivity. In the matrixial perspective, frontiers become co-poietically transgressive and limits become thresholds. A matrixial borderspace is as such a mutating co-poietic net with a co-poietic transformational potentiality, where co-creativity ("metramorphosis") might occur. As such it equally entails a particular mode of meaning and knowledge production, and is able to describe certain aspects of human symbolic experience. See Ettinger (2006).

44. See for this Gnosticism: Kusters (2009). See also Pagels (1976) and Saelid Gilhus (1984).

45. See also King (1864: 27–31), Longueville Mansel (1875: 188), Maury (1850: 484–5, 1863: 341–2), and Réville (1863).

46. Irenaeus, *Against the Heresies, 2.20.2; 1.3.3; 1.4.1* (Irenaeus of Lyons 1992), Origen (1953: 350), and Nilsson (1947: 169–72).

47. Irenaeus, *Against the Heresies*, 1.30.3–13 (Irenaeus of Lyons 1992: 95–102).

48. Irenaeus, *Against the Heresies*, 1.3.3 (Irenaeus of Lyons 1992: 28–9). The vision of the destroying Haemorrhoissa might derive from Zoroastrianism and the myth of the Great Whore who promised Ahriman, the Destructive Force of the universe, destruction and pollution in her menstruation. The sight of her blood encouraged Ahriman; he changed himself into a snake and cloaked the earth and the waters in darkness. The story is seen as reflecting a deep-rooted male fear of the feminine (Kuryluk 1991: 110; Zaehner 1975: 45–6).

49. On menstruation in the pre-medicinal world, see Shail and Howie (2005).
50. Like hematite, the amethyst, with its purple-red color, alludes to the spectrum of blood; Baert (1997: 195–210).
51. See my article devoted to this issue: Baert (2000).
52. The mountain and the cave are parts of the matriarchal cosmogony (Meissner 1925: 11).
53. The so-called *crux ansata* is said to derive from the matrix.

References

Abel, Jenötöl (ed.). 1885. *Orphica*. Lipsiae: Freytag.

Agamben, Giorgio. 1999. "On Potentiality." In Giorgio Agamben (ed.) *Potentialities: Collected Essays in Philosophy*, pp. 177–84. Stanford, CA: Stanford University Press.

Arasse, Daniel. 1996. *Le détail: Pour une histoire rapprochée de la peinture*. Paris: Flammarion.

Aretz, Erich, Embach, Michael, Persch, Martin and Roning, Franz (eds). 1996. *Der Heilige Rock zu Trier: Studien zur Geschichte und Verehrung der Tunika Christi*. Trier: Paulinus.

Asterius of Amasia. 1972. "Homily 1, PG 40, coll. 165–168." In Cyril Mango (ed.) *The Art of the Byzantine Empire 312–1453: Sources and Documents*. Englewood Cliffs, NJ: Prentice Hall.

Baert, Barbara. 1997. "The *Allegory with a Virgin*: Contributions to the Solution of an Iconographical Enigma." In Hélène Verougstraete (ed.) *Memling Studies: Proceedings of the International Colloquium (Bruges, November 10–12, 1994)*, pp. 195–210. Leuven: Peeters.

Baert, Barbara. 2000. "The Gendered Visage: Facets of the Vera Icon." *Antwerp Royal Museum Annual*: 10–43.

Baert, Barbara. 2002. "La Piscine Probatique à Jérusalem. L'eau médicinale au Moyen Age. " In Bert Cardon and Jan Van der Stock (eds) *"Als Ich Can": Liber Amicorum in Memory of Professor Dr. Maurits Smeyers*, pp. 91–129. Leuven: Peeters.

Baert, Barbara. 2004. "The Wood, The Water, and the Foot, or how the Queen of Sheba Met up with the True Cross: With Emphasis on the Northern European Iconography." *Mitteilungen für Anthropologie und Religiongeschichte* (MARG) 16: 217–78.

Baert, Barbara. 2005. "The Pool of Bethsaïda: The Cultural History of a Holy Place in Jerusalem." *Viator: Medieval and Renaissance Studies* 36: 1–22.

Baert, Barbara. 2006. "Wasserkrug und Kamm. Die Darstellung der Verena von Zurzach, ein Beispiel für neue Tendenzen in der ikonologischen Methodik." *Österreichische Zeitschrift für Volkskunde* 60(109): 35–62.

Baert, Barbara. 2007. "The Healing of the Blind Man at Siloam, Jerusalem: A Contribution to the Relationship between Holy Places and the Visual Arts in the Middle Ages." *Arte Cristiana* 838: 49–60; 839: 121–30.

Baert, Barbara. 2008. "The Twilight Zone of the *Noli me tangere*: Contributions to the History of the Motif in Western Europe (ca. 400–ca. 1000)." *Louvain Studies* 32: 255–308.

Baert, Barbara. 2009a. "'Who Touched Me and My Clothes?' The Healing of the Woman with the *Haemorrhage* (Mark 5:24b–34parr) in Early Medieval Visual Culture." *Antwerp Royal Museum Annual* (in press).

Baert, Barbara. 2009b. "In Touch with Jerusalem: *Noli me tangere*, Narrative Space and Iconic Space." *Antwerp Royal Museum Annual* (in press).

Baert, Barbara. 2009c. "The Pact between Space and Gaze: The Narrative and the Iconic in *Noli me tangere*." In Ralph Dekoninck and Agnes Guiderdoni (eds) *To Tell, to Think and to Experience Images from Theology to Rhetoric and Aesthetics in the Early Modern Period*. Leuven (in press).

Baert, Barbara. 2009d. *Navel: On the Origin of Things*. Leuven: Baert.

Barasch, Moshe. 1987. *Giotto and the Language of Gesture*. Cambridge: Cambridge University Press.

Barb, Alfons A. 1948. "St. Zacharias the Prophet and Martyr. A Study in Charms and Incantations." *Journal of the Warburg and Courtauld Institutes* 11: 35–67.

Barb, Alfons A. 1953. "Diva Matrix." *Journal of the Warburg and Courtauld Institutes* 16: 193–238.

Barta, Karen A. 1991. "Paying the Price of Paternalism." In Marie-Eloise Rosenblatt (ed.) *Where Can We Find Her?* New York: Paulist.

Beyer, Rolf. 1987. *Die Königin von Saba. Engel und Dämon. Der Mythos einer Frau*. Bergisch Gladbach: Lübbe.

Biale, David. 2007. *Blood and Belief: The Circulation of a Symbol between Jews and Christians*. Berkeley, CA: University of California Press.

Bidez, Joseph. 1913. *Philostorgius Kirchengeschichte mit dem Leben des Lucian von Antiochiën und den Fragmenten eines Arianischen Historiographen*. Leipzig: Hinrichs'sche Buchhandlung.

Bieringer, Reimund. 2005. "'Nader Mij niet': De betekenis van *mê mou haptou* in Johannes 20:17." *HTS Teologiese Studies/Theological Studies* 61: 19–43.

Bieringer, Reimund. 2006. "*Noli me tangere* and the New Testament." In Barbara Baert, Reimund Bieringer, Karlijn Demasure and Sabine Van Den Eynde (eds) *"Noli me tangere". Mary Magdalene: One Person, Many Images* (Documenta libraria, 32), pp. 16–28. Leuven: Peeters.

Bieringer, Reimund. 2008a. "'I am ascending to my Father and your Father, to my God and your God' (John 20:17): Resurrection and Ascension in the Gospel of John." In Craig R. Koester and Reimund Bieringer (eds) *The Resurrection of Jesus in the Gospel of John* (WUNT, 222), pp. 209–35. Tübingen: Mohr.

Bieringer, Reimund. 2008b. "'They have taken away my Lord': Text-Immanent Repetitions and Variations in John 20:1–18." In Gilbert Van Belle (ed.) *Repetitions and Variations in the Fourth Gospel: Style, Text, Interpretation*, pp. 609–30. Leuven: Peeters.

Boeckler, Albert. 1961. *Ikonographische studien zu den Wunderscenen der Ottonischen Malerei des Reichenau*. Munich: Bayerische Akademie der Wissenschaften.

Bovon, François. 1991. *L'évangile selon saint Luc* (Commentaire du Nouveau Testament, 3a). Geneva: Labor et Fides.

Branham, Joan R. 1999. "Frauen und blutige Räume. Menstruation und Eucharistie in Spätantike und Mittelalter." *Vorträge Warburg-Haus* 3: 129–61.

Bromley, Donald H. 2005. "The Healing of the Hemorrhaging Woman: Miracle or Magic." *Journal of Biblical Studies* 5(1): 1–20.

Bronfen, Elisabeth. 1998. *Das verknotete Subjekt: Hysterie in der Moderne*. Berlin: Volk und Welt.

Buckley, Thomas and Gottlieb, Alma. 1988. *Blood Magic: The Anthropology of Menstruation*. Berkeley, CA: University of California Press.

Bugge, Ragne. 1975. "'*Effigiem Christi, qui transis, semper honora.*' Verses Condemning the Cult of Sacred Images in Art and Literature." *Acta ad archaeologiam et artium historiam pertinentia* VI: 127–39.

Cambrensis, Giraldus. 1873. *Speculum Ecclesiae*, lib. IV, 6: Rerum Britannicarum Medii Aevi Scriptores, Vol. 21, 4, 274ff. London: Longman.

Camporesi, Piero. 1995. *Juice of Life: The Symbolic and Magic Significance of Blood*. New York: Continuum.

Clark Kee, Howard. 1983. *Miracle in the Early Christian World: A Study in*

Sociohistorical Method, pp. 162–3. New Haven, CT: Yale University Press.

Clark Kee, Howard. 1984. *Medicine, Miracle and Magic in New Testament Times*. Cambridge: Cambridge University Press.

Cohen, Shaye. 1991. "Menstruants and the Sacred in Judaism and Christianity." In Sarah B. Pomeroy (ed.) *Women's History and Ancient History*, pp. 273–300. Chapel Hill, NC: The University of North Carolina Press.

Connely, Richard H. 1929. *Didascalia Apostolorum*. Oxford: Clarendon Press.

Cotter, Wendy. 1999. *Miracles in Greco-Roman Antiquity: A Sourcebook*. London: Routledge.

Culiano, Ioan P. 1987. "Gnosticism from the Middle Ages to the Present." In Mircea Eliade (ed.) *The Encyclopedia of Religion*, pp. 574–8. New York and London: Macmillan.

Cummings, John T. 1980. "The Tassel of His Cloak: Mark, Luke, Matthew and Zechariah." In Elizabeth A. Livingstone (ed.) *Papers on the Gospels. Sixth International Congress on Biblical Studies,* Oxford, April 3–7, 1978 (Studia Biblica, 2), pp. 47–61. Sheffield: Sheffield Academic Press.

D'Angelo, Mary Rose. 1999 "Gender and Power in the Gospel of Mark: The Daughter of Jairus and the Woman with the Flow of Blood." In John C. Cavadini (ed.) *Miracles in Jewish and Christian Antiquity: Imagining Truth* (Notre Dame Studies in Theology, 3), pp. 83–109. Notre Dame: University of Notre Dame Press.

De Bruyne, Lucien. 1943. *L'imposition des mains dans l'art Chrétien ancien*. Vatican City: Pontificio istituto di archeologia cristiana.

Deckers, Johannes G., Hans R. Seeliger and Gabriele L. Mietke. 1987. "*Die Katakombe 'Santi Marcellino e Pietro.'" Repertorium der Malereien*, pp. 223–343. Vatican City and Münster: Pontificio istituto di archeologia cristiana.

Demyttenaere, Albert. 1990. "The Cloth and the Stain." In Werner Affeldt (ed.) *Fraue in Spätantike und Frühmittelalter: Lebensbedingungen, Lebensnormen, Lebensformen*, pp. 141–66. Sigmaringen: Thorbecke.

Dr Aigremont (pseudonym of Siegmar Schultze-Gallera). 1909. *Fusz- und Schuh-Symbolik und Erotik. Folkloristische und sexualwissenschaftliche Untersuchungen*. Darmstadt: Bläschke.

Dragon, Gilbert. 1991. "Holy Images and Likeness." *Dumbarton Oaks Papers* 45: 23–33.

Ebermann, Oskar. 1903. *Blut- und Wundsegen in ihrer Entwicklung dargestellt* (Palaestra, 24). Berlin: publisher unknown.

Edersheim, Alfred. 1951. *The Temple: Its Ministry and Services as They Were at the Time of Jesus Christ*. Grand Rapids, MI: Eerdmans.

Eisler, Robert. 1930. "La prétendue statue de Jésus et de l'Hémorroïsse a Panéas." *Revue archéologique* 31: 18–27.

Ettinger, Bracha L. 2006. *The Matrixial Borderspace*. Minneapolis, MN: The University of Minnesota Press.

Eusebius of Caesarea. 1955. *Histoire écclesiastique*. Trans. Gustave Bardy. Paris: Cerf.

Fabre, Pierre-Antoine. 2007. "L'image possible. Réflexions sur le défaut d'illustration dans les écrits prescriptifs et défensifs sur l'image au XVIe siècle." In Ralph Dekonick and Agnes Guiderdoni-Bruslé (eds) *Emblemata sacra. Rhétorique et herméneutique du discours sacré en images. The Rhetoric and Hermeneutics of Illustrated Sacred Discourse* (Imago Figurata Studies, 7), pp. 229–51. Turnhout: Brepols.

Fitzmyer, Joseph. 1985. *The Gospel According to Mark*. New York: Doubleday.

Fonrobert, Charlotte. 1997. "The Woman with a Blood-Flow (Mark 5:24–34) Revisited: Menstrual Laws and Jewish Culture in Christian Feminist Hermeneutics." In Craig A. Evans and James A. Sanders (eds), *Early Christian Interpretation of the Scriptures of Israel: Investigations and Proposals* (JSNTS, 148—Studies in Scripture in Early Judaism and Christianity, 5), pp. 121–40. Sheffield: Sheffield Academic Press.

Fonrobert, Charlotte. 2000. *Menstrual Purity: Rabbinic and Christian Reconstructions of Biblical Gender*. Stanford, CA: Stanford University Press.

Frazer, Margaret and Weitzman, Karl (eds). 1977. *The Age of Spirituality*. Princeton, NJ: Princeton University Press.

Fredriksen, Paula. 1995. "Did Jesus Oppose the Purity Laws?" *Bible Review* 11: 18–25, 42–7.

Gaignebet, Claude. 1976. "Véronique ou l'Image-Vraie." *Anagrom* 7: 45–70.

Gandelman, Claude. 1986. *Le regard dans le texte. Image et écriture du Quattrocento au XXe siècle*. Paris: Méridiens Klincksieck, pp. 107–12.

Gilders,William K. 2004. *Blood Ritual in the Hebrew Bible: Meaning and Power*. Baltimore, MD: Johns Hopkins University Press.

Ginzburg, Carlo. 1998. *A distance. Neuf essays sur le point de vue en histoire*. Paris: Gallimard.

Grob, Rudolf. 1967. "Art. Berühren." In Lothar Coenen (ed.) *Theologisches Begriffslexikon zum Neuen Testament*, pp. 85–6. Wuppertal: Theologischer Verlag.

Haber, Susan. 2003. "A Woman's Touch: Feminist Encounters with the Hemorrhaging Woman in Mark 5.24–34." *Journal for the Study of the New Testament* 26(2): 171–92.

Harlizius-Klück, Ellen. 2005. *Saum und Zeit: Ein Wörter-und Sachen-Buch in 496 lexikalischen Abschnitte angezettelt*. Berlin: Edition Ebersbach.

Haulotte, Edgar. 1966. *Symbolique du vêtement selon la bible*. Paris: Aubier.

Horsley, Richard A. 2001. *Hearing the Whole Story: The Politics of Plot in Mark's Gospel*. Louisville, KY: Westminster John Knox Press.

Hull, John M. 1974. *Hellenistic Magic and the Synoptic Tradition* (Studies in Biblical Theology. Second Series, 28). Napperville, IL: Allenson.

Irenaeus of Lyons. 1992. *Against the Heresies*. Trans. Dominic J. Unger (Ancient Christian Writers: The Works of the Fathers in Translation, no. 55). New York: Paulist.

Kessler, Herbert L. 1998. "Configuring the Invisible by Copying the Holy Face." In *The Holy Face and the Paradox of Representation* (Proceedings), pp. 130–51. Bologna: Nuova Alfa Editoriale.

King, Charles W. 1864. *The Gnostics and their Remains*. London: Bell and Daldy.

Knipp, David. 1998. *Christus medicus in der früchristlichen Sarkophagskulptur. Ikonographische Studien der Sepulkralkunst des späten vierten Jahrhunderts*. Leiden: Brill.

Koerner, Joseph L. 1993. *The Moment of Self-Portraiture in German Renaissance Art*. Chicago, IL, and London: University of Chicago Press.

Kötzsche, Lieselotte. 1979. In Karl Weitzmann (ed.) *Age of Spirituality. Late Antique and Early Christian Art, Third to Seventh Century*. Princeton, NJ: Princeton University Press.

Kuryluk, Ewa. 1991. *Veronica and Her Cloth: History, Symbolism, and Structure of a "True" Image*. Oxford: Blackwell.

Kusters, Liesbet. 2009. "Who is She? The Identity of the Hemorrhaging Woman and Her Wirkungsgeschichte." *Antwerp Royal Museum Annual*: 99–133

Lalleman, Pieter J. 1997. "Healing by a Mere Touch as a Christian Concept." *Tyndale Bulletin* 48(2): 335–61.

Le Quéré, François. 1997 *La sainte tunique d'Argenteuil: Histoire et examen de l'authentique tunique sans couture de Jésus-Christ*. Paris: de Guibert.

Lemarquand, Grant. 2004. *An Issue of Relevance: A Comparative Study of the Story of the Bleeding Woman (Mk 5:25–34; Mt 9:20–22; Lk 8:43–48) in North Atlantic and African Contexts*. New York: Peter Lang.

Lemay, Helen. 1990. "Women and the Literature of Obstetrics and Gynecology." In Joel T. Rosenthal (ed.) *Medieval Women and Medieval History*, pp. 189–209. Athens, GA: University of Georgia Press.

Levine, Amy-Jill. 1996. "Discharging Responsibility: Matthean Jesus, Biblical Law and Hemorrhaging Woman." In David R. Bauer and Mark A. Powell (eds) *Treasure New and Old: Recent Contributions to Matthean Studies*, pp. 379–97. Atlanta, GA: Scholars Press.

Leyerle, Blake. 2001. "Blood is Seed." *Journal of Religion* 81(1): 26–48.

Longueville Mansel, Henry. 1875. *The Gnostic Heresies of the First and Second Centuries*. London: John Murray.

Lurker, Manfred. 1991. "Art. Gefäss." In *Wörterbuch der Symbolik*, pp. 232–3. Stuttgart: Kröner.

Mancinelli, Fabrizio. 1987. *De Romeinse catacomben en de wortels van het Christendom*. Florence: Scala.

Marcus, Joel. 2000. *Mark 1–8: A New Translation with Introduction and Commentary*. New York: Doubleday.

Marshall, Christopher D. 1989. *Faith as a Theme in Mark's Narrative* (Society for New Testament Studies. Monograph Series, 64), pp. 106–8. Cambridge: Cambridge University Press.

Mathews, Thomas F. 1993. The *Clash of Gods: A Reinterpretation of Early Christian Art*. Princeton, NJ: Princeton University Press.

Maury, Alfred. 1850. "Lettre à M. Raoul Rochette sur l'étymologie du nom de Véronique." *Revue Archéologique* 7: 484–95.

Maury, Alfred. 1863. *Croyances et légendes de l'antiquité*. Paris: Cerf.

May, Eric E. 1952. "'For Power Went Forth from Him...' (Luke 6:19)." *Catholic Biblical Quarterly* 14(2): 93–103.

McCracken, Peggy. 2003. *The Curse of Eve, The Wound of the Hero: Blood, Gender and Medieval Literature* (The Middle Ages Series). Philadelphia, PA: University of Pennsylvania Press.

Meier, Christel. 1977. *Gemma Spiritualis. Methode und Gebrauch der Edelsteinallegorese vom frühen Christentum bis ins 18. Jahrhundert*. Munich: Wilhelm Fink Verlag.

Meissner, Bruno. 1925. *Babylonien und Assyrien*, Vol. 1. Heidelberg: Winter.

Melzer-Keller, Helga. 1997. *Jesus und die Frauen. Eine*

Verhältnisbestimmung nach den synoptischen Überlieferungen (Herders biblische Studien, 14). Freiburg: Herder.

Metternich, Ulrike. 2000. *"Sie sagte ihm die ganze Wahrheit": Die Erzählung von der "Blutflüssigen" feministisch gedeuted.* Mainz: Matthias Grünewald Verlag.

Morello, Giovanni and Wolf, Gerhard. 2000. *Il volto di Cristo* (exhibition catalog), Rome, Palazzo delle Esposizioni. Milan: Electa.

Murray, Margaret A. 1911. "Note on the SA Sign." *Man* 11(73): 113–77.

Nancy, Jean-Luc. 2000. *Corpus.* Paris: Editions Métailié.

Nancy, Jean-Luc. 2008. *Corpus.* Trans. Richard A. Rand. New York: Fordham University Press.

Nauerth, Claudia. 1983. "Heilungswunder in der frühchristlichen Kunst." In Herbert Beck and Peter C. Bol (eds) *Spätantike und frühes Christentum,* pp. 339–46. Frankfurt am Main: Liebieghaus Museum alter Plastik.

Neumann, Erich. 1956. *Die Grosse Mutter.* Zurich: Rhein Verlag.

Neusner, Jacob. 1973. *The Idea of Purity in Ancient Judaism.* Leiden: Brill.

Nilsson, Martin P. 1947. "Sophia-Prunikos." *Eranos. Acta Philologica Suecana* 45(1–2): 169–72.

Oppel, Dagmar. 1995. *Heilsam erzählen—erzählend heilen. Die Heilung der Blutflüssigen und die Erweckung der Jairustochter in Mk 5,21–43 als Beispiel markinischer Erzählfertigkeit* (Bonner biblische Beiträge, 102). Weinheim: Beltz.

Origen. 1953. *Contra Celsum, VI, 35.* Trans. Henry Chadwick. Cambridge: Cambridge University Press.

Ouaknin, Marc-Alain. 1994. *Lire aux éclats: Éloge de la caresse.* Paris: Seuil.

Noga-Banai, Galit. 2008. *The Trophies of the Martyrs: An Art Historical Study of Early Christian Silver Reliquaries.* Oxford: Oxford University Press.

Pagels, Elaine H. 1976. "What Became of God the Mother? Conflicting Images of God in Early Christianity." *Signs* 2(2): 293–303.

Paracelsus. 1922–33. *"De matrice."* In Karl Sudhoff (ed.) *Sämtliche Werke,* Vol. 9. Munich: Oldenbourg.

Parish Sanders, Ed. 1990. *Jewish Law from Jesus to the Mishnah.* London: SCM.

Parish Sanders, Ed. 1998. *Judaism. Practice and Belief. 63BCE–66CE.* London: SCM.

Pettazzoni, Raffaele. 1921. "Le origini della testa di Medusa." *Bolletino d'arte* 1: 491–509.

Plaskow, Judith. 1993. "Antijudaism in Feminist Christian Interpretation." In Elisabeth Schüssler Fiorenza (ed.) *Searching the Scriptures: A Feminist Introduction,* pp. 117–29. New York: Crossroad.

Pointon, Marcia. 1986. "Interior Portraits: Women, Physiology and the Male Artist." *Feminist Review* 22: 5–22.

Pradel, Fritz. 1907. *Griechische und Süditalische Gebete, Beschwörungen und Rezepte des Mittelalters.* Giessen: Töpelmann.

Réville, A. 1863. "La Véronique, une sainte Gnostique." *Le Lien* 28.

Rhoads, David. 1994. "Jesus and the Syrophoenician Woman: A Narrative-Critical Study." *Journal of the American Academy of Religion* 62(2): 343–75.

Robertson, Archibald T. and Perschbacher, Wesley J. 2004. *Word Pictures of the New Testament,* Vol. 1. *Matthew and Mark.* Grand Rapids, MI: Kregel.

Ronig, Franz J. 2005. "Erläuterungen zu den Miniaturen des Egbert Codex." In Sif D. Dornheim, Doris Oltrogge, Robert Fuchs, Franz J. Ronig and Günther Franz (eds) *Der Egbert Codex. Das Leben Jesu. Ein Höhepunkt der Buchmalerei vor 1000 Jahren.* Stuttgart: Theiss.

Rubin, Nissan and Kosman, Admiel. 1997. "The Clothing of the Primordial Adam as a Symbol of Apocalyptic Time in the Midrashic Sources." *Harvard Theological Review* 90(2): 155–74.

Ryken, Leland. 1998. "Art. Garments." In Leland Ryken and James C. Wilhoit (eds) *Dictionary of Biblical Imagery,* pp. 317–20. Downers Grove, IL: InterVarsity Press.

Saelid Gilhus, Ingvild. 1984. "Gnosticism: A Study in Liminal Symbolism." *Numen* 1(31): 106–28.

Schiel, Hubert. 1960. *Codex Egberti der Stadtbibliothek Trier.* Basel: Alkuin.

Schiller, Gertrud. 1971. *Iconography of Christian Art,* Vol. 1. *Christ's Incarnation, Childhood, Baptism, Temptation, Transfiguration, Works and Miracles.* Trans. Janet Seligman. London: Humphries.

Schüssler Fiorenza, Elisabeth. 1994. *In Memory of Her: A Feminist Theological Reconstruction of Christian Origins*. New York: Crossroad.

Selvidge, Marla J. 1984. "Mark 5:25–34 and Leviticus: A Reaction to Restrictive Purity Regulations." *Journal of Biblical Literature* 104(4): 619–23.

Selvidge, Marla J. 1990. *Woman, Cult, and Miracle Recital: A Redactional Critical Investigation on Mark 5:24–34*. London and Toronto: Bucknell University Press.

Shail, Andrew and Howie, Gillian (eds). 2005. *Menstruation: A Cultural History*. New York: Palgrave Macmillan.

Sidgwick, Emma. 2009. Tactiliteit en potentialiteit in het motief van de Haemorrhoissa (Mc 5:24–34parr). "Cultureel-antropologische verkenningen." *Antwerp Royal Museum Annual*: 135–61.

Söding, Thomas. 1985. *Glaube bei Markus. Glaube an das Evangelium, Gebetsglaube und Wunderglaube im Kontext der markinischen Basileiatheologie und* Christologie (Stuttgarter Biblische Beiträge, 12). Stuttgart: Katholisches Bibelwerk.

Spier, Jeffrey. 1993. "Medieval Byzantine Magical Amulets and Their Tradition." *Journal of the Warburg and Courtauld Institutes* 56: 25–62.

Struthers Malbon, Elizabeth. 1984. "The Jesus of Mark and the Sea of Galilee." *Journal of Biblical Literature* 103(3): 363–77.

Struthers Malbon, Elizabeth. 1992. "Narrative Criticism: How Does the Story Mean?" In Janice C. Anderson and Stephen D. Moore (eds) *Mark and Method: New Approaches in Biblical Studies*, pp. 23–47. Minneapolis, MN: Fortress Press.

Swidler, Leonard. 1971. "Jesus was a Feminist." *Catholic World* 212: 177–83.

Taylor Gench, Frances. 2004. *Back to the Well: Women's Encounters with Jesus in the Gospels*. Louisville, KY, and London: Westminster John Knox Press.

Teteriatnikov, Natalia. 1995. "The Place of the Nun Melania (the Lady of the Mongols) in the Deesis Inner Narthex of Chora, Constantinoplei." *Cahiers Archéologiques* 43: 163–80.

Theissen, Gerd. 1983. *The Miracle Stories of the Early Christian Tradition*. Philadelphia, PA: Fortress Press.

Trummer, Peter. 1991. *Die Blutende Frau. Wunderheilung im Neuen Testament*. Freiburg and Vienna: Herder.

Twelftree, Graham H. 1999. *Jesus the Miracle Worker: A Historical and Theological Study*. Downers Grove, IL: InterVarsity Press.

Underwood, Paul. 1966. *The Kariye Djami*. Vol. 1: *Historical Introduction and Description of the Mosaics and Frescoes*. New York: Pantheon Books.

van der Loos, Hendrik. 1965. *The Miracles of Jesus* (Novum Testamentum Supplementum, 9). Leiden: Brill.

van Loo, Sofie (ed.). 2007. *Gorge(l): Oppression and Relief in Art*. Antwerp: Royal Museum of Fine Art.

Vandenbroeck, Paul. 2000. *Azetta. Berbervrouwen en hun kunst,* (exhibition catalog), Brussels, Palais des Beaux-Arts. Ghent: Ludion.

Vandenbroeck, Paul. 2009. "The Energetics of an Unknowable Body." In Paul Vandenbroeck and Gerard Rooiakkers (ed.) *Backlit Heaven: Power and Devotion,* pp. 174–205. Mechelen: LAMOT.

Veith, Ilza. 1965. *Hysteria: The History of a Disease.* Chicago, IL: University of Chicago Press.

Verhoeven, Cornelius W. M. 1956. *"Symboliek van de voet."* Doctoral dissertation, Assen.

Viscuso, Patrick. 1991. "Purity and Sexual Defilement in Late Byzantine Theology." *Orientalia Christiana Periodica* 57: 399–408.

von Dobschütz, Ernst. 1899. *Christusbilder. Untersuchungen zur christlichen Legenden.* Leipzig: Hinrichs.

Vööbus, Arthur. 1979. *The Didascalia Apostolorum in Syriac* (CSCO, 401–2, 407–8). Leuven: Peeters.

Waagenvoort, Hans. 1957. "Contactus." *Revue de Archéologie Chrétienne* 3: 404–42.

Walker Bynum, Caroline. 1992. *Fragmentation and Redemption: Essays on Gender and the Human Body in Medieval Religion.* New York: Zone Books.

Walker Bynum, Caroline. 2007. *Wonderful Blood: Theology and Practice in Late Medieval Northern Germany and Beyond.* Philadelphia, PA: University of Pennsylvania Press.

Warwick, Alexandra and Cavallaro, Dani. 2001. *Fashioning the Frame: Boundaries, Dress and the Body.* Oxford: Berg Publishers.

Watson, Carolyn J. 1981. "The Program of the Brescia Casket." *Gesta* 20(2): 283–98.

Whitekettle, Richard W. 1996. "Levitical Thought and the Female Reproductive Cycle: Wombs, Wellsprings, and the Primeval World." *Vetus Testamentum* 46(3): 376–91.

Wilkinson, John. 1998. *The Bible and Healing: A Medical and Theological Commentary.* Grand Rapids, MI: Eerdmans.

Wilson, Ian. 1991. *Holy Faces, Secret Places: The Quest for Jesus' True Likeness.* London: Doubleday.

Wolf, Gerhard. 1998. "From Mandylion to Veronica: Picturing the 'Disembodied' Face and Disseminating the True Image of Christ in the Latin West." In *The Holy Face and the Paradox of Representation* (Proceedings), pp. 153–79. Bologna: Bologna: Nuova Alfa Editoriale.

Wolf, Gerhard. 2002. *Schleier und Spiegel. Traditionen des Christusbildes und die Bildkonzepte der Renaissance.* Munich: Wilhelm Fink.

Wolf, Gerhard, Bozzo, Colette D. and Calderoni Masetto, Anna R. 2004. *Mandylion: Intorno al Sacro Volto, da Bisanzio a Genova.* Milan: Skira Editore.

Wood, Charles T. 1981. "The Doctor's Dilemma: Sin, Salvation, and the Menstrual Cycle in Medieval Thought." *Speculum* 56(4): 710–27.

Zaehner, Robert C. 1975. *The Teachings of the Magi.* London: Sheldon Press.

Textile Commerce and Songket Creativity: The Role of Heritage Entrepreneurs in Contemporary Gold-thread Weaving in Sumatra

Abstract

The weaving and marketing of "traditional textiles" today in places like Indonesia often involve not only thread markets, weavers, customers, and ritual wearers but also savvy heritage entrepreneurs who conceptualize and promote such cloths as important icons of ethnic peoplehood. This article explores the key role of such heritage entrepreneurs in the contemporary design and production of metal-thread songket textiles in Pandai Sikek, West Sumatra, and in Palembang in South Sumatra. Drawing on fieldwork interviews with indigenous cloth heritage brokers and the weavers they mentor, the article focuses on specific cloths to illuminate dimensions of songket's striking resiliency and artistic creativity today. Long a cloth of the Asian marketplace, Sumatran songket today is emerging as a textile of social class mobility, a transition yielding new weaving styles "within tradition."

Keywords: Sumatran songkets, heritage entrepreneurs, anthropology of textiles

SUSAN RODGERS

Susan Rodgers is Professor of Anthropology, College of the Holy Cross, Worcester, MA. Her PhD is from the University of Chicago (1978) and she studies the politics and aesthetics of Indonesia's minority literatures and arts. Her publications include translations of Sumatran novels and *Gold Cloths of Sumatra*, with Anne and John Summerfield (Iris and B. Gerald Cantor Gallery and KITLV Press, 2007).

Textile, Volume 9, Issue 3, pp. 352–371
DOI: 10.2752/175183511x13173703491153
Reprints available directly from the Publishers.
Photocopying permitted by licence only.
© 2011 Berg. Printed in the United Kingdom.

Textile Commerce and Songket Creativity: The Role of Heritage Entrepreneurs in Contemporary Gold-thread Weaving in Sumatra

In Sumatra, peninsular Malaysia, and some coastal regions of Kalimantan, Sulawesi, and several eastern Indonesian islands, songket cloths are luxury textiles hand-loomed in silk or fine cotton to which gold- or silver-wrapped threads have been woven in across the weft, as decorative elements. These supplementary wefts form elaborate geometric, botanical, or more rarely, animal figure designs. The songket textile would have complete integrity without the additional, shiny yarns woven in from right to left across the weaver's front, as she sits at her loom and manipulates songket's distinctive arrays of bamboo or wooden pattern sticks. Some songket production is done on frame looms (for instance, in West Sumatra, by Minangkabau weavers; Figure 1) while other songkets are woven on back-tension looms. The latter is the case in Palembang and Jambi in southern Sumatra (Figure 2). Songket weaving has ties to Malay world court arts of the 1600s and 1700s and quite possibly to the cloth trade that coursed between Muslim north Indian textile centers and trade entrepôt cities in Sumatra and the Malay peninsula between the 1500s and 1700s.[1]

This international trade history has made songket far different from the handspun cotton ritual cloths that are often caught up in geographically narrower village cycles of exchange found in relatively smaller-scale Southeast Asian ethnic minority societies (e.g. Toba's *ulos* textiles, which use ikat dyeing techniques). Cosmopolitan to their core, songkets have historically been cloths of the Muslim marketplace and the sultan's palace in places like West and South Sumatra. These regions are classic hubs of Asian interregional trade in luxury goods. They are foci also of Malay World networks of Muslim business families close to fabric production.

Perhaps even more fundamental to West and South Sumatran songket's nature as a textile type is the fact that these radiant cloths in effect embody their trade histories in their physical and technical nature. That is, for one thing, they are made of "the very stuff" of foreign lands and trans-island travel (silk, gold thread, typically not indigenous to Sumatra). Songkets wrapped into ritual clothing for brides and lineage chiefs help wearers to claim high social status. But, these songkets

Figure 1
A Minangkabau songket weaver in Pandai Sikek, West Sumatra, working at a frame loom. Photo: Susan Rodgers.

also give the owner the cachet of ambitious personal migrations beyond a home village, which might well not produce its own gold-thread cloth. In addition, songket's motif repertoires are intensely cosmopolitan and some are near pan-Southeast Asian. Some songket weaving traditions also rely on features such as representations of fern tendrils and interlocking geometrics—design formats that resonate with interregional Muslim world aesthetics. All of these motif conventions imbue songkets with a woven-in level of design sophistication and an allure of "the foreign."

Additionally—and centrally to Sumatran songket's aesthetic— when new luxury threads from places like Japanese factories first become available in an old songket weaving region such as Palembang, craftswomen often start using these new materials with dispatch, according to recent interviews. They do this to keep their "traditional" songket textiles as up-to-date, and as lustrous, as possible. Weavers also forthrightly seek saleability for their newly woven, contemporary yet "old style" songkets.

This article explores some current-day instances of this signature songket characteristic of embodying complex trade histories within songket textiles' overt material culture form and format. I look particularly at connections between current-day songket design and the styles of

Figure 2
Southern Sumatran weavers using
back-tension looms to weave songket.
Photo: Susan Rodgers.

commerce pursued today by the Muslim textile business families that commission and sell newly woven yet supposedly old-style songkets. Prominent among these families are older women and men who act as heritage entrepreneurs. These are persons who are in the business of selling ethnic peoplehood traditions via this specific expensive, labor-intensive, glittery cloth. These heritage entrepreneurs interact intensely with weavers to produce today's finest songkets. Leaving the heritage entrepreneurs out of the equation in songket textile scholarship can limit our view of these remarkable cloths and their equally notable resiliency over time. Studying Sumatran songket's precise heritage entrepreneurship economic settings today adds dimension to approaches that focus primarily on the weaver and the customer.

Pandai Sikek and 30 Ilir as Commercial Settings for Selling Tradition

The locales for this study are Pandai Sikek in the Minangkabau highlands of West Sumatra and Palembang, South Sumatra. These are two historically distinct and geographically distant songket weaving centers within this large island. The Minangkabau people are distinct from the Palembang Malays. However, the regions nonetheless show similarities in terms of pronounced linkages between textile commerce (as superintended and shaped by heritage entrepreneurs) and songket creativity. Songket textiles in both areas stand in some contrast to many other Indonesian indigenous arts, which have sometimes been characterized as fragile (for instance, Taylor 1994). West and South Sumatran songket handloom industries are thriving today, with hundreds of weavers, dozens of songket shops, and wide clienteles in Indonesia and Malaysia. Sumatran songket's artistic robustness today, I contend, links in significant part to the business savvy of heritage entrepreneurs, who often hail from old weaving families.

A specific element of Sumatran Muslim weaving textile commerce is key here: the ways in which lavish, enticing emporia built in architectural styles that mimic old traditional houses are used to sell especially upscale songket and to simultaneously "sell heritage." These emporia often constitute the home base for the most prominent Minangkabau or Palembang Malay heritage brokers of songket cloths. In West Sumatra these business people have women weavers working for them on commission, on home looms. In Palembang, heritage entrepreneurs hire young women to weave songket on looms set up on small shop floors. These are sometimes located in pavilions adjacent to the weaving house emporia.

The weaving house entrepreneurs often see themselves as songket *designers*. In interviews, they often use that English word within Indonesian language discourse (interviews for this study were in Indonesian). They also talk about conveying great old weaving traditions from the distant past into the present; they claim to be the prime movers in maintaining high standards of technical weaving excellence. They talk of time, art— and management.

To investigate contemporary songket creativity in relation to weaving house commercial settings and the heritage entrepreneurs who manage these emporia, I shall examine four individual songket cloths that emerged from prominent weaving houses/ shops in the West Sumatran craft village of Pandai Sikek and from the South Sumatran 30 Ilir weaving neighborhood of Palembang. All four textiles embody Indonesia's present-day definition of songkets as heritage wear. That is, these songkets work in the public imagination as emblems of "tradisi," ethnic tradition, within Indonesia's swirl of many imagined ethnic traditions.

These four cloths were all brokered over the last fifteen years and were consciously designed by the Minangkabau and Palembang Malay entrepreneurs who run these textile business establishments. These enterprises are ones that might be termed "traditional weaving houses-gone-modern" (Figure 3). That is, these are family-run heritage cloth enterprises whose principals (often prosperous haji pilgrims) seek broad new customer bases among Indonesia's emerging middle classes. This sector of the population tends to be university educated and has professional and managerial jobs. The emerging Indonesian middle class includes Minangkabau and Palembang Malay families whose antecedents often had little direct access to songket luxury wear (for they did not come from the traditional nobility). These families now have the disposable income to purchase luxury songkets, for ceremonial occasions. This precise business setting of brokers playing to new middle class clienteles has shaped the warps and wefts of these cloths in inventive ways. This builds on songket's centuries of life as cloths of Southeast Asian commerce. Today's songkets are now, in part, textiles of social class mobilities.

As noted, the songket-selling enterprises I explore here are located in shops whose architecture consciously evokes the center rooms of old Minangkabau or old Palembang Malay "traditional houses." That is, the emporia mimic the old multigenerational structures that have elaborate roofs and many side rooms, clear public signs of inherited wealth in Sumatra. In Minangkabau, these side chambers were once for the households of

Figure 3
The interior of Pak Zainal's songket shop, Palembang, South Sumatra. The architecture invokes the styles of a
traditional Palembang Malay house. Photo: Susan Rodgers.

daughters; in the songket emporia, the side chambers display special types of gold cloth for sale.

These emporia, though, point far beyond so-called traditional house architecture. The shops also draw inspiration on the marketing plane from the so-called heritage floors of large downtown Jakarta department stores such as Pasaraya and Sarinah. From these tradisi-drenched shops (with all their allusions as well to malls and upscale shopping forays on weekends) emanate (ironically enough) songket cloths that are both democratic in customer appeal and stylistically innovative. This textile innovation occurs in terms of motif band juxtaposition patterns, motif construction, thread type, dye choice, and overall compositional format. This is a songket tradition, in other words, that fosters considerable creativity and change at the level of the individual cloth. All of this takes place alongside designers', weavers', wearers', and sellers' heavy rhetorics about remaining fervently faithful to songket weaving tradition. Songket entrepreneur/designers operate in this precise cultural space, as they seek greater sales for their quite pricey luxury textiles.

I draw on fieldwork interviews from 1995 to 2008[2] with weavers and songket heritage entrepreneurs from both locales mentioned above. Their narratives about songket production and songket histories allow examination of some of the connections between these four textiles' concrete Minangkabau, Palembang Malay, and wider Indonesian business settings and the weavers' minute, specific design decisions in motif execution and in selecting threads of a certain hue and twist. After discussion of theoretical background we can move on to the four songkets.

Brokering Heritage Textiles: Comparative Contexts

In "Introduction: Textile Economies of Value, Sanctity, and Transnationalism," anthropologists Walter Little and Patricia McAnany note that textiles that index a folk culture have long been entangled with trans-border, transcommunity political economies. They write,

> *The nineteenth-century industrialization of the textile industry . . . illustrates the broad economic, social, and political spheres through which textiles interpenetrate. Such embedded economic processes (Polyani 1944) are global in scope and entangle many different cultural groups with capitalism. Just as English textile mill workers were thrust into global capitalism, so nineteenth-century Navajo wool producers and weavers (M'Closkey 2002) suffered and continue to suffer . . . within an economic system that exploits and marginalizes them while, at the same time, valorizing*

> *and replicating transnationally their new creations.* (Little and McAnany 2011: 1–2)

Little and McAnany go on to note (p. 2) that such "transglobal processes have been part of the matrix of cloth for centuries." Sumatran songket's most recent upswing in creativity, documented below in the case studies, has been shaped by the translocal political economies of heritage goods that Little and McAnany describe. Yet, this textile form has emerged in vibrant health. Several articles in *Cross-Stitching Textile Economies with Value, Sanctity, and Transnationalism* (Little and McAnany 2011) help fill in the comparative context.

In "Exchange without Brokers: Weaver-Client Relationships in Senegal," Laura L. Cochrane (2011) notes that West African textile art is most typically marketed to national and international clients by long-distance Senegalese brokers, who shuttle between arts-producing villages, urban markets and shops, and national culture venues such as arts exhibitions. Christopher Steiner has documented ways some of these brokers penetrate international markets (1994). Cochrane's work focuses on weavers who interact directly with clients: an unusual circumstance and one that indicates much agency. In Sumatran songket production, broker/ entrepreneurs loom especially large and in effect seek to control "their" weavers' artistic lives.

In "Creativity, Place, and Commodities: The Making of Public Economies in Andean Apparel Industries," Rudi Colloredo-Mansfield *et al.* (2011)

document how brokers today sometimes interact intensively with international technologies of weaving (computerized looms, for instance) and with international customer bases for folk art. Individual weavers become somewhat eclipsed within such transactions. Sumatran songket entrepreneurs are less overbearing and more solicitous of weavers. They broker tradition more softly, eschewing overt commercialization of craftwork. We can now turn to the four songkets to see how Minangkabau and Palembang Malay heritage entrepreneurs shape production and imagine consumption.

An Aqua Shoulder Cloth from the Pusako Weaving House, Pandai Sikek

The bright aqua and blue songket shoulder cloth shown in Figure 4, worked with lavish amounts of silver-wrapped threads, has a large customer fan-ship. It is of a type that has become so popular over the last five years or so that the Pusako Weaving House has a hard time keeping it and its matching sarong in stock. Both pieces are designed for Minangkabau women's ceremonial wear, in *adat* (in "inherited custom"). This matching set was woven by a Pandai Sikek woman at her home loom (a frame loom) and then given to this weaving house on consignment for sale. The two-piece set in aqua can be purchased for about US$70.00 to $80.00. This is a significant amount of money in West Sumatra. This shoulder cloth was targeted by its designer (the Pusako Weaving House's proprietress, eighty-four-year-old

Figure 4
A selendang or shoulder cloth woven
in 2005 in Pandai Sikek, West Sumatra.
Silk and polyester blend, sliver-
wrapped thread. Iris and B. Gerald
Cantor Art Gallery Study Collection,
Holy Cross. Photo: Frank Graham and
Roger Hankins, for Cantor Art Gallery,
Holy Cross.

retired weaver Ibu Sanuar) for
the middle of the fine songket
purchasing market. That is, this
songket was intended for sale to
women such as schoolteachers
and school principals, bank staff,
or lower-level government workers
who save up for months to make a
major ceremonial wear purchase
for an adat occasion such as a
relative's wedding.

Another important part of Ibu
Sanuar's perceived market for this
piece is the Minangkabau middle
aged woman who has a veritable
collection of songket ensembles
for adat wear and who wishes to
keep up with each new variation
in design format and color. Some
of these Minangkabau women live
in the *rantau*, the region beyond
the home areas of West Sumatra;

they reside in upscale middle- and upper-middle-class neighborhoods of South Jakarta, for instance. Having fine songkets from Pandai Sikek to wear to ceremonies marked for Minangkabau identity is especially important to well-off Minangkabau women living in the emphatically national Indonesian and international city of Jakarta, far from their families' Sumatran home region.

Several features of this songket make it "the latest thing" while it still maintains an overall aura of traditionalism and forthright Minangkabau-ness. In basic composition, this shoulder cloth is unexceptional in its materials for current-day Minangkabau songkets: silk and polyester blend threads, imported silver-wrapped threads. But, the aqua color is distinctive. Ibu Sanuar and other retired weavers now working as shop entrepreneurs/songket designers use commercial dye color choice to push the market along, in a consumer culture sense. They work to assure that last year's color in songket thread will quickly be perceived as passé. The designer/ entrepreneurs do make sure, however, that certain classic colors, such as deep red, remain more than acceptable, year to year. Deep red tones are seen as *ur*-traditional. Indeed, they are, in terms of songket's dye color history.

Since the 1995 songket season, for instance (an accurate term, in these high-end songket ateliers), the popular thread colors among weavers and buyers have run through a series of changes: bright, clear reds; intense purples and blues; "crème," a creamy white that was especially in demand

for songkets totally covered in motif bands; then the lime green popular several seasons ago on international fashion catwalks; then these aqua textiles. A monied Minangkabau woman collector of songkets for ceremonial wear does attempt to buy one of every major new color type.

Another feature of this shoulder cloth that speaks to both its design nature and its larger marketing setting is the fact that it has a series of clear-cut, easily discernible bands of motifs. These are set off from each other by plain bands, in contrasting colors. This shouts out "new style" in Pandai Sikek heritage cloth houses. There are antique songket antecedents for this thin-band style but this is not a commonly encountered design format today.

Here, Ibu Sanuar has asked the home weaver to use a linked pair of very large double cut-off diamond motifs at the base of the songket. This is a forthright allusion to the old motifs that her Pusako Weaving House is famous for bringing back into Minangkabau weaving. This motif is this weaver's take on the *salapah* tobacco box design (Summerfield and Sutan Madjo Indo 1999: 180, 184). Going up the songket, though, the weaver has employed several plainer bands with very simple, flat diamond motifs: this is a new motif. Additionally, the motif bands are borrowed from several *different* old weaving villages' weaving histories. The broad selvedge bands, however, are Pandai Sikek's own diagnostic ones. Weavers and songket designers/heritage entrepreneurs do not distain this approach of combining different

settlements' motifs. Rather than being seen as a jumble (which might well have been the case in the past), songkets of this sort are praised by today's weavers, brokers, and customers for their creativity. But this is a creativity within tradition, as the individual motifs (of whatever village origin) should ideally remain stable. This is Ibu Sanuar's view for her weaving establishment. She is a strict taskmaster with weavers in this regard. If they exercise creativity at the level of individual motifs she will reject their textiles and not offer them for sale.

When a Minangkabau woman wears a songket ensemble of this type to an adat ceremony, Minangkabau ethnic identity is invoked in roundabout ways. Whereas through the 1920s, at least, most major Minangkabau adat domains had their own distinctive songket motif repertoires and modes of tying them into women's headdresses (A. Summerfield 1999: 105–71), today a woman wearing a cloth such as this as a shoulder covering is announcing a more generalized, even genericized Minangkabau-wide identity. Others also can and do buy these songkets. One example is the following. Many Minangkabau families migrated to Negeri Sembilan, Malaysia, in the nineteenth century. The wealthier among them now sometimes come on songket shopping tourism trips to Pandai Sikek in Indonesia, to purchase higher-quality yet more sensibly priced songkets than they can get in Malaysian markets. Malaysia has protectionist legislation to protect their own songket industries, although canny shoppers like Ibu Sanuar's Malaysian customers can circumvent these laws. A cut-rate airline flies direct from Kuala Lumpur to Padang, West Sumatra, and heritage shoppers can take vans or taxis directly to Pandai Sikek from the airport for their shopping forays.

A growing, emergent, pan-Minangkabau traditionalist identity, inflected for middle-class socioeconomic status, is indexed here in this decidedly unsubtle songket, with its cascade of motif bands, its jaunty coloration—and its clear fashion world shelf-life limitations and verve.

No-gold-thread Songkets

Figure 5 shows a Pandai Sikek-woven, contemporary songket that has the unusual feature of totally lacking metal-wrapped thread. Rather, the supplementary wefts are silk. This yields a type

Figure 5
A West Sumatran songket woven without gold-wrapped threads. The supplementary wefts are silk. Iris and B. Gerald Cantor Art Gallery Study Collection, Holy Cross. Photo: Frank Graham and Roger Hankins, for Cantor Art Gallery, Holy Cross.

of songket with increasing allure for upscale Minangkabau women customers. The reason? The textile can be laundered.

The major songket brokers who work with Pandai Sikek weavers today are exploring this sector of supplementary weft weaving with much interest. The market setting at least in part is as follows. Songket sarongs, headties, and shoulder cloths made with metal-wrapped thread are used in ceremonial dress in Minangkabau's hot, humid climate. The metallic textiles get sweated down after hours of ceremonial processions in the sun. Yet, if the songkets are washed the metal-wrapped threads will dull. So, consumers are left on the horns of a dilemma. They have made a major purchase for ceremonial fabric, in part to assure a glorious public appearance at rituals, yet one ritual in the tropics can threaten to harm the future visual appeal of their songket wear. The sheer impracticality of metallic-thread songkets has driven some Minangkabau upscale women buyers toward a cheaper and certainly more washable alternative for some ceremonial occasions: *busana Muslim*, Muslim fashion wear, sewn from regular fabric (that is, not made of metal-thread songket). The no-gold-threads songkets are being strategically developed in competition with these cheaper, washable non-songket Muslim wear fashions.

In West Sumatra today, busana Muslim garb typically consists of a long, tight skirt and a matching long fitted jacket (a *baju kurung*). They are made of polyester fabrics and are accompanied by an embroidered Muslim-style head covering (a *jilbab*). These baju kurung sets mimic Malaysian styles for Muslim women and offer a marked contrast to Indonesia's official national dress for women (a batik sarong, a lacy kebaya jacket, a light and in fact optional headscarf, and a shoulder cloth that matches the sarong or kebaya in some way). Busana Muslim wear varies in price from about US$30.00 at the low end to costumes costing hundreds of dollars. But a Minangkabau woman can feel quite well turned out in a Muslim wear, busana Muslim ensemble that comes in under $50.00—a situation unlikely with songket-based garb.

Virtually the entire Minangkabau population is Muslim and songket itself is strongly associated with Sumatra's Islamic communities. For instance, by contrast, the predominantly Christian Toba Batak rarely use songket. However, when a Minangkabau woman dons a busana Muslim polyester outfit for a formal occasion she is clearly claiming religious piety, in terms of modest dress and symbolic references to resurgent Muslim conservatism at the national Indonesian level.[3] Over the last thirty years, the busana Muslim fashion sector for Minangkabau women has grown exponentially. Many shops in Bukittinggi that sell songket also offer busana Muslim fabrics, ready to be taken to a seamstress.

Since even the fanciest busana Muslim sets are completely washable, and since some are reasonably priced, to boot, songket entrepreneurs in Pandai Sikek have begun to worry about erosion of their customer base. A few are fighting back, by commissioning

these no-gold-thread songkets. Textiles like that shown in Figure 5 employ standard songket motif registers and certainly look like regular songket, yet they can be washed and used for years. They also have an especially supple hand. They are expensive, however. This derives from the fact that only the most experienced weavers can produce them, since silk supplementary wefts are slippery and harder to work with than metallic threads.

It is notable that Ibu Sanuar herself, the proprietress of Pusako Weaving House, has not yet offered any of these no-metal songkets for sale. She has asked several of her best weavers to experiment with the style and has given them the materials for that. So far, though, their songkets have not measured up to her exacting standards. The cloth in Figure 5 was sold by another broker.

What this jockeying for market share between songket dealers and purveyors of polyester Muslim fashions portends for the future of Minangkabau songket handloom industries is hard to predict. So far, metal-thread songket sales are still robust at emporia like Ibu Sanuar's. She seems to be experimenting with the new sorts of all-silk songkets around the edges of her enterprise, so as to carefully protect her business's health on into the future.

Ibu Us's Fantasy on a Pitalah Headcloth

This cloth (Figure 6), which is boldly metallic, evokes the extravagantly long headtie lengths of silk worked with gold thread associated with the magnificent weavings from

Figure 6
A recently woven songket for wall
display, Tanjung, Pandai Sikek, West
Sumatra. Woven by Ibu Us. Linen-
polyester blend, gold-wrapped thread.
Iris and B. Gerald Cantor Art Gallery
Study Collection, Holy Cross. Photo:
Frank Graham and Roger Hankins, for
Cantor Art Gallery, Holy Cross.

Pitalah Village. This is a highland Minangkabau settlement whose women weavers produced exquisitely fine-detailed ceremonial belts and headdress cloths for the Minangkabau sumptuary goods market in the mid-1800s (J. Summerfield 1999: 91, Fig. 5.34 shows a fine example). Weaving in Pitalah is defunct today: the dozen or so great Minangkabau weaving villages of the late 1700s to 1920s now typically do other or no craftwork today. The town of Silungkang yields some good songkets today, via more mechanized processes. But there is only one West Sumatran village that produces very high quality songket weavings today on handlooms: Pandai Sikek. A few looms are active in Batu Sangkar and in Payakumbuh, but Pandai Sikek dominates the fine weaving handloom industry of West Sumatra. Women weavers in Pandai Sikek often trace explicit genealogical connections to Pitalah weavers of the late 1700s. This cloth was woven by forty-year-old Ibu Us in her home on her frame loom in the Tanjung neighborhood of Pandai Sikek. The textile employs commercially dyed linen-polyester blend threads, chosen at the thread market by Ibu Us herself, and gold-wrapped yarns. The textile combines a plain weave technique with numerous thick motif bands worked in supplementary weft, in metal-wrapped threads (some from Japanese factories). This textile is described in detail in Rodgers *et al.* (2007: 7, 60–1, 93–4).

This songket was born in the heads of a well-known, locally well-regarded husband–wife team of Minangkabau art and antique dealers who run a shop in the city of Bukittinggi directed to international connoisseurs of Sumatran arts. They are Sutan Madjo Indo and his wife Ibu H. Elly Azhar. Both are haj pilgrims, long active in Muslim charity work for disadvantaged schoolchildren from their home village. Their shop, Toko Aladdin, once had arts customers in some number in the 1980s and 1990s, when there was still a sizeable expatriate community of American oil company families living in nearby Pekan Baru. But, since the Bali bombing of 2002, Indonesia's perceived Islamist terrorist threat, and the post-1997 persistent downturn in sectors of the Indonesian economy, few well-heeled international residents of Sumatra or Singapore come through Bukittinggi intent on antiquing. Backpackers still do appear in small numbers but they rarely have the cash or interest to purchase this particular shop's old daggers, betel serving sets, or old songkets. The latter can easily run to hundreds or even thousands of dollars. So, by the late 1990s, the proprietors began to diversify. Ibu Elly moved into buying and selling Muslim fancy wear embroidered clothing and Muslim headscarf ensembles for Bukittinggi women, and Sutan Madjo Indo moved into commissioning relatively affordable, newly woven songket "replicas" of old textile types that Minangkabau families had once owned in their house treasures but had long since sold for cash to antiques dealers like himself.

In conceptualizing his new product shown in Figure 6, Sutan Madjo Indo imagined (he tells me) that current-day families from Pitalah might sometimes regret the fact that they no longer have their fine nineteenth-century silk songket headties, shoulder cloths, and sarongs to wear to such events as weddings and the installations of lineage chiefs. Many such persons now live in Sumatran cities or in Jakarta but they still need copious amounts of ceremonial songket wear. So, Sutan Madjo Indo's marketing strategy continued, if he could commission replacement Pitalah songkets from an especially expert weaver, the resultant cloths could stand in for the lost, actual antique textiles, all the while being newer looking, not to mention having fewer mouse and moth holes. Another of his business aims was to foster a market in displayable textile wall hangings: new yet old-style songkets for Minangkabau urban émigrés living in multiethnic Indonesian cities such as Jakarta.

In the late 1990s Sutan Madjo Indo and Ibu Elly zeroed in on the relatively young, brilliant weaver Ibu Us as their partner in this endeavor. She was not working exclusively for any of Pandai Sikek's traditional weaving house brokerages. A freelancer, she was selling her fine cloths from her home and by special commission to wealthy women such as governors' wives. The Madjo Indo's provided Ibu Us with startup money and fragments of old Pitalah songkets as motif templates. They suggested that she weave the old forms with great precision and that she choose commercially dyed fine threads whose hue and diameters would allow her to evoke the subtle old cloths. They tried to procure the very best gold-wrapped threads for her to work with. Pandai Sikek weavers today aver that the quality of gold thread was much higher before about the 1920s. Today's gold thread, they say, is rough and stiff (on gold thread in West Sumatran songket, see Indictor 1999). Us's new songkets were not to be sold to customers as "fakes," by any means. To Minangkabau's extremely knowledgeable songket clienteles, her new cloths would be immediately recognizable as what they were—late 1990s and early 2000s versions of old cloths commissioned and woven in contemporary Indonesia.

The business arrangement was that the Madjo Indo's would sell Ibu Us's songkets of this sort to either songket-less Pitalah families, or to ex-pats traveling through Bukittinggi with a good deal of money to spend. Neither sort of customer materialized, so Sutan Madjo Indo approached a Minangkabau business acquaintance of his who runs several arts and antiques boutiques in upscale international hotels in Jakarta and in Bali. Sutan Madjo Indo suggested that Ibu Us's "Pitalah" songkets might be displayed there for sale to hotel guests. When these hotel locales also failed to yield paying customers, of any sort, Ibu Us and the Madjo Indos mutually agreed that the set of nine replica antique-style songkets that she had woven in this craft niche would be her last. Ibu Us has since turned her attention to weaving more mundane though still quite beautiful sorts of songkets, for ceremonial wear as sarongs and shoulder cloths. These textiles resemble the standard run of Pandai Sikek songkets available in the heritage shops.

This so-called replacement songket has some features distinctive to its gestation from negotiations between Ibu Us and her would-be brokers. First, it is notably more luxurious-looking than the types of songkets generally sold today in heritage shops. Its length is almost overwhelmed with a glinting, onrushing march of metal-thread motif bands, whereas a more standard shoulder cloth would have less goldwork, and fewer, less densely packed motif bands. Secondly, this is not a Pitalah replica in any sense of the word, in a motif composition sense. The motif bands are mixed together in ways that no nineteenth-century Pitalah weaver would countenance (although some of the individual pairs of bands are more conventional). For instance, the botanical forms in rows at the tops of the patterned sections take a basic Pitalah weaving idea but go on to reshape them, making them wider and longer. Ibu Us has also framed her weaving with the elaborate selvedge bands of motifs that are often used today by most all Pandai Sikek weavers. These selvedge bands make a songket immediately recognizable as a Pandai Sikek product among Minangkabau songket connoisseurs.

What has resulted is more in the nature of a riff, or a fantasy, on Pitalah weaving conventions, than any replica per se. Ibu Us seems at base to have created a tangible icon of tradition, but one, as noted, that has no popular fan base—except for me, since I have purchased several of Ibu Us's songkets of this sort for the

Southeast Asian textile study collection in the College of the Holy Cross's Iris and B. Gerald Cantor Art Gallery. We displayed this textile as part of the 2007 exhibition, *Gold Cloths of Sumatra: Indonesia's Songkets from Ceremony to Commodity* at Holy Cross's Iris and B. Gerald Cantor Art Gallery. This is a profoundly globalized songket of much artistic energy.

A Beaded and Bangled Shoulder Cloth from the Fikri Koleksi Family, Palembang

This quite different songket (Figure 7, sold in 2005 and woven shortly before) uses cotton blend threads, gold-wrapped threads, and a positive profusion of little glass beads and fancy fringe. In fact, this is a songket that is all but weighted down by its bells and whistles—a design element that the Palembang Malay family that sold it were quite forthright about. In 2006 I asked why the textile had so ornate a buzz of gold and glass in and around it. A clerk replied: that is what is popular now. A fairly expensive songket for one that uses all cotton threads (as opposed to silk), this shoulder cloth and its matched sarong (equally bedazzling) cost the rupiah equivalent of US$105.00.

Figure 7
A shoulder cloth or selendang from Pak Zainal's shop, Palembang, South Sumatra. 2005. Cotton-blend, gold-wrapped thread, glass beads and fringe. Photo: Frank Graham and Roger Hankins, for Cantor Art Gallery, Holy Cross.

The main Fikri Koleski shop in 30 Ilir includes a weaving pavilion out back staffed by young village women, most in their late teens and early twenties. Songket work depends on sharp eyesight so weavers are relatively young. This particular cloth was bought at this large extended family's other shop in 30 Ilir, their old ancestral home, literally a Malay house. The eldest son of the family runs this establishment and uses it as his Palembang home base; his other shop is in an upscale neighborhood of downtown Jakarta. There, he sells a variety of newly woven songkets along with some antique textiles, to select clients.

This black, gold, and glass songket was conceptualized and woven in considerable contradistinction to Palembang's stereotypic, classic songket type: that is, one called *kain limar*, woven all in silk, colored a deep, rich red, with a center field of delicate silk ikat in subtle pinks and oranges. These standard Palembang kain limars need no bells and whistles to impress: their subtle ikat work is overwhelming in its beauty and softness and these cloths' gold thread supplementary work is superb. But Palembang weaving houses typically are not as innovative as Ibu Sanuar's establishment is in Pandai Sikek, in the sense that the Palembang designers are not mining old motif registers to constantly reinvigorate their motif bands repertoires. Palembang songkets typically hew pretty closely to a predictable array of florals, swirls, boxes, and stars. The Fikri Koleksi family does encourage weavers to follow the latest trends in thread color (one

current favorite: "baby pink" [*sic*]). To create a distinctive profile in a crowded Palembang songket scene, then, entrepreneurs like the eldest son here sometimes move beyond the songket per se, to glass bead fringes and embellishments.

This songket entrepreneur—better termed, impresario—is Pak Zainal. He is something of a Jakarta celebrity and has ambitions to do for Sumatran songket what batik artists such as the late Iwan Tirta once did for Javanese batik—that is, to make it the stuff of fashion legend. Beyond selling self-consciously traditionalist kain limar songkets, and fine songkets of the bells and whistles variety, Pak Zainal is also attempting to refashion Palembang songket into cloth that can be confected into tony evening gowns and shoulder wraps, for Jakarta's fashionistas.

But, in an irony of songket materiality, that translation of textile type is proving resistant. Songket cloth cannot be cut, to be made into pieces following a dress pattern to be sewn together into a clothing item such as a fitted gown. Songket is, after all, metal cloth.

Conclusion

Marketing tradition and marketing art can be a hazardous business in a predominantly Muslim nation like Indonesia anxious about public religious identities and worried also about alternative modernities available to all Southeast Asian publics through mass media. But, marketing songkets as heritage art and as wearable, adat-attuned luxury goods can also have considerable benefits to weavers, wearers, and sellers, as examination of these four songkets has shown. Songket weaving is expanding vigorously into new cultural arenas for all these groups, as they claim Minangkabau and Palembang Malay historical authenticities all the while they continue to celebrate, and attain, markedly high levels of weaving excellence. The concrete business context of weaving house emporia leaders working as songket heritage entrepreneurs has fostered this sector of Indonesian textile creativity and resilience.

Acknowledgments

I am grateful for the sponsorship of the Lembaga Ilmu Pengetahuan Indonesia and also Andalas University, Padang, for the early stages of this field study. Holy Cross's Iris and B. Gerald Cantor Art Gallery and Holy Cross's Research and Publication Committee provided generous support for the fieldwork. My deepest thanks go to the weavers and weaving house leaders in Pandai Sikek (Ibu Sanuar and her son Pak Yan), 30 Ilir in Palembang (the family of Haji Agus Hussein Rahmat at the Fikri Koleksi), and also to Ibu Masfar, director of the extraordinary women's craft cooperative in Kotogadang, West Sumatra, an establishment called Amai Setia.

Notes

1. A key source on Minangkabau songket is Summerfield and Summerfield (1999). For broader discussion of contemporary songket in West Sumatra, South Sumatra, see Rodgers *et al.* (2007). Kartiwa (1986) is a good introduction to songkets in Indonesia in general. Selvanayagam (1990) and especially Mohamad (1996) provide valuable comparative material on Malaysian songket industries. East Balinese weavers have a distinctive songket home industry. The scholarship on Indonesian textiles in more general terms is large. Especially rewarding are Maxwell (2003, 2004); classic sources are Gittinger (1989, 1990). For broader geographical coverage to wider Southeast Asia, see especially Fraser-Lu (1990). Forshee's study of Sumba ikats on the move through various commercial contexts (2001) provides illuminating comparative context for studies of the marketing of contemporary Sumatran songkets.

2. This fieldwork has entailed a total of ten months of residence in Bukittinggi, West Sumatra since 1995 and three shorter visits to Palembang, Jambi, and Bengkulu, as well as interviews with émigré Minangkabau in Jakarta. All interviews were conducted in bahasa Indonesia.

3. There is a valuable anthropological literature on issues of Muslim fashion, specifically the jilbab, in contemporary Indonesia. Core sources are Smith-Hefner (2007), Brenner (1996), and Jones (2007, 2010).

References

Brenner, Suzanne. 1996. "Reconstructing Self and Society: Javanese Muslim Women and the 'Veil.'" *American Ethnologist* 23(4): 673–97.

Cochrane, Laura L. 2011. "Exchange without Brokers: Weaver-Client Relationships in Senegal." In W. Little and P. McAnany (eds) *Cross-Stitching Textile Economies with Value, Sanctity, and Transnationalism*. Lanham, MD: AltaMira Press (in press).

Colloredo-Mansfield, J. Antrosio, Munuela, A. and Jones, E. 2011. "Creativity, Place, and Commodities: The Making of Public Economies in Andean Apparel Industries." In W. Little and P. McAnany (eds) *Cross-Stitching Textile Economies with Value, Sanctity, and Transnationalism*. Lanham, MD: AltaMira Press (in press).

Forshee, Jill. 2001. *Between the Folds: Stories of Cloth, Lives, and Travels from Sumba*. Honolulu, HA: University of Hawai'i Press.

Fraser-Lu, Sylvia. 1990. *Handwoven Textiles of South-East Asia*, 2nd edn. Singapore: Oxford University Press.

Gittinger, Mattiebelle (ed.). 1989. *To Speak with Cloth: Studies in Indonesian Textiles*. Los Angeles, CA: Museum of Cultural History, University of California–Los Angeles.

Gittinger, Mattiebelle. 1990. *Splendid Symbols: Textiles and Tradition in Indonesia*, 2nd edn. Singapore: Oxford University Press.

Indictor, Norman. 1999. "Metallic Threads in Minangkabau Textiles." In A. Summerfield and J. Summerfield (eds) *Walk in Splendor: Ceremonial Dress and* the Minangkabau, pp. 225–38. Los Angeles: UCLA Fowler Museum of Cultural History, Textile Series 4.

Jones, Carla. 2007. "Fashion and Faith in Urban Indonesia." *Fashion Theory* 11(2/3): 211–32.

Jones, Carla. 2010. "Materializing Piety: Gendered Anxieties about Faithful Concumption in Contemporary Urban Indonesia." *American Ethnologist* 37(4): 617–37.

Kartiwa, Suwati. 1986. *Kain Songket Indonesia* (*Songket Weaving in Indonesia*). Jakarta: Djambatan.

Little, Walter and McAnany, Patricia A. (eds). 2011. *Cross-Stitching Textile Economies with Value, Sanctity, and Transnationalism.* Lanham, MD: AltaMira Press (in press).

Little, Walter and McAnany, Patricia. 2011. "Introduction: Textile Economies of Value, Sanctity, and Transnationalism." In W. Little and P. McAnany (eds) *Cross-Stitching Textile Economies with Value, Sanctity, and Transnationalism.* Lanham, MD: AltaMira Press (in press).

M'Closkey, Kathy. 2002. *Swept under the Rug: A Hidden History of Navajo Weaving.* Albuquerque, NM: University of New Mexico Press.

Maxwell, Robyn. 2003. *Textiles of Southeast Asia: Tradition, Trade, and Transformation,* rev. edn. Singapore: Periplus.

Maxwell, Robyn. 2004. *Sari to Sarong: 500 Years of Indian and Indonesian Textile Exchange.* Canberra: National Gallery of Indonesia.

Mohamad, Maznah. 1996. *The Malay Handloom Weavers: A Study of the Rise and Decline of Traditional Manufacture.* Singapore: Institute of Southeast Asian Studies.

Polyani, Karl. 1944. *The Great Transformation: The Political and Economic Origin of our Time.* Boston, MA: Beacon Press.

Rodgers, Susan, Summerfield, Anne and Summerfield, John. 2007. *Gold Cloths of Sumatra: Indonesia's Songkets from Ceremony to Commodity.* Worcester, MA: Iris and B. Gerald Cantor Art Gallery; Leiden, the Netherlands: KITLV Press.

Selvanayagam, Grace Impam. 1990. *Songket—Malaysia's Woven Treasure.* Singapore: Oxford University Press.

Smith-Hefner, Nancy. 2007. "Javanese Women and the Veil in Post-Soeharto Indonesia." *Journal of Asian Studies* 66(2): 389–420.

Steiner, Christopher B. 1994. *African Art in Transit.* Cambridge: Cambridge University Press.

Summerfield, Anne. 1999. "Women's Ceremonial Dress and Related Ceremonial Textiles." In A. Summerfield and J. Summerfield (eds) *Walk in Splendor: Ceremonial Dress and the Minangkabau,* pp. 105–70. Los Angeles, CA: UCLA Fowler Museum of Cultural History, Textile Series 4.

Summerfield, John. 1999. "Men's Ceremonial Dress." In A. Summerfield and J. Summerfield (eds) *Walk in Splendor: Ceremonial Dress and the Minangkabau,* pp. 79–104. Los Angeles, CA: UCLA Fowler Museum of Cultural History, Textile Series 4.

Summerfield, Anne and Summerfield, John (eds). 1999. *Walk in Splendor: Ceremonial Dress and the Minangkabau*. Los Angeles, CA: UCLA Fowler Museum of Cultural History, Textile Series 4.

Summerfield, Anne and Sutan Madjo Indo, H. A. 1999 "Motifs and their Meanings." In A. Summerfield and J. Summerfield (eds) *Walk in Splendor: Ceremonial Dress and the Minangkabau*, pp. 171–200. Los Angeles, CA: UCLA Fowler Museum of Cultural History, Textile Series 4.

Taylor, Paul M. (ed.). 1994. *Fragile Traditions: Indonesian Art in Jeopardy*. Honolulu, HA: University of Hawai'i Press.

Curls and Culture: Hairstyles in British Cinema 1930–80

Abstract

This article analyzes the cultural meanings of hair from an anthropological point of view, and argues that head hair is used to create *distinction*. The performance of hairstyles speaks about the class, leisure, and cultural competence of the wearer. The article sets up a number of terminologies for hairstyle analysis: symmetry/intensity/composition/ innovation/cultural quotation/ technology.

The article then looks at female hairstyles in British cinema. It problematizes the idea that hairdressers could be ascribed with agency in production, and looks at the meaning of female hairstyles in the 1930s. It then considers hairstyles in films aimed at female audiences in the 1940s. The article analyzes the narrative function of female hairstyles in films of later periods, particularly the 1970s, and argues that hair operated as a symbol of desire, fear, and innovation in a range of film texts. Archival and textual analysis of *Barry Lyndon* suggests that under certain rare circumstances, the meaning of hairstyles could break loose from the intentions of the director or producer.

Keywords: hairstyles, cultural history, British cinema

SUE HARPER

Sue Harper is Emeritus Professor of Film History at the University of Portsmouth. She has published widely on British cinema. Sue was Principal Investigator of a major AHRC research project on British cinema of the 1970s. She and Justin Smith have completed the book of the project, *British Film Culture of the 1970s: the Boundaries of Pleasure,* which will be published by the University of Edinburgh Press in 2011.

Textile, Volume 9, Issue 3, pp. 372–383
DOI: 10.2752/175183511X13173703491199
Reprints available directly from the Publishers.
Photocopying permitted by licence only.

Curls and Culture: Hairstyles in British Cinema 1930–80

Hair is never neutral. It is one of those bodily manifestations that human beings use to impose order and structure on the world. In this, it is precisely like those other parts of the body—penis, vulva, breasts—which change with the process of time or desire, and which can be scarified, cut, concealed, displayed or adorned. Such body parts are unlike bodily products such as excrement, urine, semen, and menstrual blood. With these, their production is necessary to the healthy regulation of the organism. Of course, such products take on a *symbolic* value in the life of the individual psyche or social groups—they can be polluting or (more unusually) confer a benison—but in themselves they do not change.

Hairs, like nails, grow and change, and can thus be subject to a whole range of procedures. Both hair and nails ensue from live tissue, but are in themselves dead, and in that regard are symbolically extremely potent. What do human beings use hair *for*? Primarily to create *distinction*, of course. The way the hair is arranged speaks volumes about the type of labor, time or adornment that its owner can afford, and is thus crucial in establishing their status. In non-Western tribal cultures, hair is a vital tool at the disposal of their members, and can signal their marriageability, their age, the number of their livestock, and their menstrual status. In many periods in Western culture, hairstyles were a signal of class ascription: the peasants' hairstyle (and dress) had to be fitting to their station. And hairstyles are a site of conspicuous consumption too, in that one needed leisure in order to be able to *perform* fashionable hairstyles.

A useful comparison in this regard is of women's fashionable hairstyles in mid- to late-nineteenth-century Britain, and in the early days after the French Revolution. In the earlier period, ships in full sail, colorful birds, and mountain landscapes adorned the heads of those who, in consequence, needed help getting into and out of a carriage. This kind of deliberate and reckless stylization was difficult to achieve and to wear. It relied on structure, which heightened the coiffeur with a sort of wicker scaffolding, and on ornamentation, which involved the addition of objects unrelated to the ensemble (Cross 2008: 15–26). It was deployed exclusively by the aristocracy and the *beau-monde*. In the later period, with the violent fall of the *ancien regime* in France, the dominant fashionable hairstyle in Britain followed that of its neighbors': a sort of poodle cut, with short curls (which might fancifully be seen as evoking "natural" pubic hair). This style had the occasional ribbon binding it up, as a reference to the style of the "moral" Roman Republic recalled from classical sculpture. So within a relatively short historical period, female hairdressing could evoke politics, class, and consumption.

We can safely assume, therefore, that hair can be used to carry a range of meanings. The ideas of anthropologist Mary Douglas are useful here. In *Purity and Danger* and other books, she argues, *pacē* Levi-Strauss and other structuralist thinkers, that humans construct their mythological systems as a way of signaling, in an implicit way, those systems of thought which articulate the society. Successful groups are those in which the structuring oppositions—those sets of opposites between which the group is poised—are easily internalized by everyone. Such oppositions often have to do with ideas of pollution and purity, and are a way of ratifying those taboos which make a social group coherent (Douglas 1966, 1973). Hair is a signifying system like any other social practice, such as clothing or body language, and as such it functions as part of a mythological system. Hair can express, in its form and reputation, socially conservative or innovatory practices; its fashions and styles reside deep in the culture, where they may evoke an aroma of power, desire or distaste.

If we are to proceed from a descriptive to an analytical account of the social function of hairstyles, we need precise terminology for them. We might think about them in terms of the following categories, in order to allow us to make discriminations between different periods and practices:

- *symmetry*. Are the two halves of the stylistic arrangement the same? A balanced hairstyle evokes a sense of coherence: an off-center one shows a tolerance of disorder. Asymmetry can throw an ensemble into disarray, and intimate that there is room for individual maneuver around the diktats of high fashion.
- *intensity*. Is it a "busy" hair performance, or a suave, seemingly-effortless one? Does it explicitly display the artistry of the hairdresser, or does it try to conceal the labor that went into the completed style? If there are a number of competing tensions or trends within the ensemble, how are they articulated, and are any of them given prominence?
- *composition*. What is the relationship of the details (curls, waves) to the overall design? Are the decorations organically integrated into the whole, or are they autonomous within the ensemble? And what is the role of accident or incompetence in the composition?
- *innovation*. Does the hairstyle conform to the dominant fashions of its period? Or does it constitute a break with established practice? How does fashion change in hairstyling? Does it proceed, like other cultural form, via a series of leads and lags? Can the notion of the dominant, the emergent, and residual, which work so well for other cultural forms, be fruitfully applied to hairstyling?
- *cultural quotation*. What historical or cultural references are being made within a hairstyle, and why? Are the quotations uniformly from one source, or are they more eclectic?
- *technology*. Does the hairstyle deploy the newest technology? The cultural historian of hairstyles needs to be aware of what was technically possible at specific periods. We need to know, for example, when Marcel Waves began, when the permanent wave became *de rigueur* and what its effects were upon hair texture, and we need to know when new dyes were inaugurated, in order to judge a hairstyle appropriately. Hollywood studios took a great deal of trouble with the hair color of stars such as Greer Garson and Lucille Ball; for the former, the red hair made her less dull, and for the latter, it provided an excuse for her wackiness.
- *competence*. We need to be able to recognize when the person working on the cut and curl showed some artistry, or when they were a ham-fisted hacker. Many a good concept has been ruined by an over-enthusiastic scissorhands.

Thus equipped, we can go on to think about the role of hair in cultural history, in film history, and in British film history in particular. I want to confine myself here to female head hair; to look at male and body hair would be too big a task. In general, in painting and sculpture, female hair has two functions. Firstly, in portraits and memorials of various sorts, the artist endeavors to *show it as it was*: to render the surface of the person and to show us how they dressed their hair. But secondly, in artworks that are not biographical memorials, female hair can operate on a symbolic level and can provide an index of desire. Rembrandt, Renoir, and Rubens are masters in this regard, and their female hairstyles symbolize a moist, luxuriant otherness that stimulates the viewer. That distinction works for representations of hair in paintings and sculptures, but not for those in the cinema. Paintings in the pre-twentieth-century period are originally made for individual patrons (usually male) who pay for them. Films, on the other hand, are designed for mass consumption. They produce profit for the makers by making pleasure for the audience, at least half of which is female.

Hairstyles in cinema therefore have, in any period, both to stimulate and appease that mass (and gendered) market. There is an intimate relationship between hairstyles on the screen and the behavior of the female viewer in the 1930s and 1940s, and this is quite well documented through the studios' publicity material—those brochures sent to cinema managers advising them how to market a film. Almost the whole of Gainsborough's material for *Madonna of the Seven Moons* (1944) is made up of pictures of the hairstyles from the film "which any girl can do herself." Not only is the viewer invited to admire, but instructed how to emulate. Photographs from the film are reproduced alongside pen-and-ink sketches of the procedures— bobby-pins, curlers, and all—which will transform the female viewer into a simulacrum of the star.[1]

Hairdressing in films that address the female audience is intimately related to fashion and implies a familiarity with it. In the 1930s, for example, a magazine called *Film Fashionland*, which was extremely popular with female audiences in Britain, suggested appropriate hairstyles to match dress ensembles that they could see in the films. In the same period, *Film Weekly* and *Picturegoer* paid minute attention to the coiffeurs of British and American women stars. It looks as though studios monitored the magazines of the hairdressing trade pretty carefully. But, from the evidence contained within studio publicity materials, the viewers were not encouraged to emulate the furthest reaches of innovation, but the middle ground: that which women could afford without looking silly. Film magazines, female stars, and studio publicity materials gave major *currency* to new hairstyles, and operated as a sort of conduit between high hair fashion and the audience: but they rarely instigated innovation themselves, since the films that they were marketing did not do that either.

But before we get down to film analysis in detail, let us problematize the idea of writing a history of hairdressing in the film industry. It is impossible to argue that hairdressers in British cinema had the power to impose their designs on the overall fashion "look" of the film. They simply did not have the institutional clout, up until the 1970s. Before then, hairstylists were subordinated to costume and makeup departments in British cinema, and were so low down in the pecking order that no one has thought it worthwhile to document them. From the 1930s to the mid-1960s, each studio had its own hairdressing department, which was quite minimally staffed and whose personnel did not come from West End salons. It was very different in America. There were "star" hairdressers in America who became attached to certain actresses and thus attained a degree of power. Sydney Guilaroff was imported by Joan Crawford and had a major career at MGM, and other star coiffeurs were Eugene Soulieman and Frederic Fekkal. All these were stylists who originated outside the studio, who made it to the inner circle, and who had a distinct input into the visual language of the films in which their stars appeared. This happened very rarely in British cinema. In the 1950s, Raymond ("Mr Teasy-Weasy") was a very influential stylist, whom Diana Dors imported into the Rank Organisation as her personal stylist, but he was an exception. Perhaps because of his very difference, Raymond was lampooned by Norman Wisdom in *On the Beat* (1962). Raymond was certainly effeminate in his repertoire of gestures, according to the television evidence, but the film satirizes these in a merciless manner. When the studio system broke down in Britain in the early 1970s, this had enormous consequences for the allocation of artistic autonomy in various areas of production, including hairdressing, as I shall show later.

So it is problematical to ascribe agency to hairdressers in British cinema in the earlier period, since they had so little power during the production process. Whose responsibility were the hairstyles,

in that case? It depends on who had the last say in the production process. If it was the director, then the hairstyles would follow his diktat and would provide evidence of his authorship: if the producer was the determinant in the last instance, then ditto. If (unusually) the production was design-led, then the designer may have given some autonomy to the hairdresser. But the latter is an unusual case.

British cinema, from the coming of sound in the late 1920s to the late 1950s, was not a director's cinema, with the possible exception of pre-Hollywood Hitchcock, Lean, and Powell and Pressburger. In the 1930s and 1940s, it was a producer's cinema. It was the producer, or the studio, who set a number of creative practices in motion, and who set up negotiations between them. It is from this point of view that we should interpret the hairstyles in *The Private Life of Henry VIII* (1933), one of the first British films to succeed on world screens. Producer Alexander Korda, who also directed the film, had a very high opinion of the mass audience's cultural competence (Korda 1937). He thought that costume film should make a living bridge between the past and the present, and was not interested in historical accuracy, since its academic aroma was likely to alienate mass audiences.[2] The hairstyles in *Henry VIII* are clearly meant to intensify empathy between the modern female viewers and the films' historical protagonists. With one exception, Henry's wives sport modern hairstyles, which accord ill with the Tudor headdresses. In one scene, Jane Seymour interrupts a political

discussion between the King and his courtiers with a question about the wedding apparel: "Listen, darling, this is really important—shall it be the chaplet or the coif?" Both are pulled off to display a modern hairstyle, flat over the crown and with impeccably permed curls at the nape. Using the terminology devised above, this is a symmetrical coiffure, in which the curls are not particularly well integrated into the overall design. All the women in the film, who either are or aspire to be the King's paramour, wear the same style. Except one: Anne of Cleves, the ugly wife whom the King cannot bed. Her hairstyle is composed of thick braids, gracelessly hanging over her ears, and the cultural quotation is of a no-nonsense Germanic utilitarianism. By implication, Anne's lack of *savoir faire* in the hair department is as big a handicap in the amatory stakes as her love of garlic. The implication of the hair designs in *Henry VIII* is that modernity in hair is coterminous with desirability. The dresses may give a nod in the direction of historical probity, but to do the same with hairstyles is a step too far.

Extensive debates took place among critics and viewers about the film's historical accuracy, which focused on its clothes, manners, and hairstyles (Chapman 2005: 27–30; Harper 1994: 22–3). Other costume films of the 1930s stimulated debates about the function of history and the benefits or otherwise of accurate representation, and some of them even attracted opprobrium for their hairstyles.[3] In general, 1930s costume films followed Korda's lead, in that they made some cursory nods in the direction of accuracy in the clothes, but made the female protagonists' hairstyles, gaze pattern, and body language resolutely modern.

There is a marked homogeneity in hair performance in 1930s British films. In the modern-dress films, female hairstyles uniformly carry information about the characters' sexual mores. In the Travers farces, such as *Turkey Time* (1933) and *A Cup of Kindness* (1934), the hairstyles offer a method of distinguishing between sexual sophisticates (who wear flat Marcel-style waves round the face, with a flat "cap" over the crown and with rows of curls beneath) and old-fashioned prudes (who wear a bun). But other genres deploy the same stylistic language. In Jessie Matthews musicals such as *It's Love Again* (1937), this version of the Marcel is the standard choice for willful or liberated females, as it is in melodramas like *The Squeaker* (1937). The hegemony of this hairstyle can be established by looking at the hair of Nina Mae McKinney, a black American actress who plays the African princess in *Sanders of the River* (1935). As an Afro-American, her "natural" hair would be curly or braided. But she sports the flattest, shiniest Marcel-with-curls of them all, in which her glistening hair is slicked close to the scalp. Young women only appear without curls or waves in 1930s British cinema when they are impersonating men, as in *Me and Marlborough* (1935), *First a Girl* (1935) or *Wings of the Morning* (1937).

During World War II, the Ministry of Information (MoI) was in control of all media—radio, theater, journalism, posters, and the cinema. Its Films Division organized shorts and documentaries, and it had its own production arm, the Crown Film Unit: it operated a kind of carrot-and-stick approach towards the commercial film industry, rewarding those who complied with its preferences, and refusing aid to those who were recalcitrant. Doubtless following Government behests, the Films Division had a particular steer that it wished to give to film, and a particular message about the necessity for women to be virtuous and self-sacrificing (Harper 2000: 31– 5). In dress, demeanor, and hairstyles, the women in films promulgated or made by the MoI are modest and controlled. The "good" women keep their hair under wraps, either literally or metaphorically. Indeed, one of the MoI posters exhorted women, after Russia had joined the Allies, to "Wear a Headscarf—your Russian Sister does!" This of course ignored the fact that the Russian women wore the scarf for *cultural* reasons, whereas the MoI wanted it to be worn for *industrial* and safety reasons. Anyway, what is noticeable in all the MoI-backed films is the predominance of the "Victory Roll" among the "good" women. This unflattering style, in which the hair is scraped back, rolled round a sort of sausage and pinned in place without compensatory curls or waves, signals duty and the avoidance of erotic distraction. Otherwise, ratified women cover their heads, as in *Millions Like Us* (1943). The farming women in the Hutterite community in *The 49th Parallel* (1941), the only commercial film that the MoI actually paid money for, wear modest headscarves and sit in a regimented row with their eyes downcast.

It is clear that MoI films presented the ideal hairstyles as utilitarian and without decoration, and that women's clear desire for expressivity in hair matters was viewed negatively by official film-makers. During World War II, fashionable hairstyles were longer and the curls looser, possibly as a consequence of two shortages: of the materials for permanent waves, and of the leisure time required for the procedure. Usually a "perm" could take a whole afternoon or longer, and with the exigencies of work and queuing, that now became impossible. So the tight curls and controlled waves of the 1930s became inconvenient. Official advice was that women should keep their hair short and straight for work, and there were such shortages of hair-dye that its used was discouraged (Lant 1991: 105). Women had to conform to the latter, but they do not appear to have heeded the former, if we adduce magazine evidence.

But more was at issue. The new looser, freer curls could be interpreted as a compensatory response by women to the limitations they experienced on other fronts. They might have dress coupons and Utility clothes, and they might be conscripted, but at least, outside work hours, they could use their hairstyles to express that freedom and

independence that many were experiencing for the first time. Now, what is clear is that, unlike the proscriptive MoI films, commercial films aimed at female audiences *recognized* this, and they deployed hair symbolism in suggestive ways. Korda's films are a case in point. In *The Thief of Bagdad* (1941) and *Lady Hamilton* (1942), both aimed at women, the heroines sport the same hairstyle, although separated by centuries and continents: hair that swoops and swings, with lush, lustrous sausage-like curls.

One commercial studio deployed the new hairstyles in a most extreme way. Gainsborough Studios made a series of costume melodramas that completely dominated the box office from 1942 to 1946, and that were aimed exclusively at women. These were films that showed women on the high road to their own desires, which they gained by guile and deviance. The gender and popular bias of the films meant that they had very low critical status. Gainsborough was organized along classic Taylorist lines, where the management led from the top and ran a series of *insulated* and *relatively autonomous* areas of production, which had little to do with each other—décor, costume, scripts, and so on. The result was that the films manifested a series of discourses that seem to contradict each other. This narrative chaos clearly had no effect on the films' popularity: I have suggested that the audience was able to pick its way with great competence through the discursive jungle that the films present (Harper 1994: 119–35).

The narrative of hair carried some weight within Gainsborough films, and it clearly functions on a subliminal level. But this was the result of managerial carelessness rather than deliberation. The costume designers and the art directors were left to work unhindered, and we must assume that the hairdressers were as well. In *Madonna of the Seven Moons* (1944), the heroine has a split personality due to a childhood sexual trauma: one self is the repressed Catholic wife of a rich banker, and the other is a wild, passionate gypsy. The heroine's transformation from one self to the other is signaled by the changes in her hair, clothes, and jewelry. In a key scene, she awakes from her repressed self and looks in the mirror. Her first act is to pull her hair free from the restraining bonds and to puff it up, making it look luxuriant. The shaking loose of her locks ushers the heroine into the world of desire and expressivity from which her former repressions have barred her. And in *The Wicked Lady* (1945), the aristocratic heroine becomes a highwayman in order to satisfy her secret, wilder nature. All her hairstyles have whorls or topknots, and the structure of the hairstyles issue from a hidden center. When her rival enters the arena as a sexual competitor, she too begins to sport such hairstyles. They deploy a sort of vaginal symbolism that probably spoke volumes to the female audience, who time after time praised the protagonists' hairstyle and clothing, and wished to emulate them. This expressive use of hair, as an index of unbridled female sexuality, runs right through the Gainsborough melodramas.

In the 1950s, it was the distributors who were the chief agents in British cinema, rather than the production companies, and the chain of command—with the big companies of Rank, ABPC, and British Lion—was so complex that dissident representations and discourses could sometimes wriggle through. Hence the rise of Diana Dors in this period. She appeared in films that were imperfectly controlled by the distribution companies that financed them, and her style of outrageous *jouissance* is signaled as much by her hair as her sumptuous figure. Bleached to white-blonde (*à la* Monroe and Jayne Mansfield), Dors' hair signals a challenge to social taboo all on its own. Mountainous, richly waved, her coiffeur displays an astonishing amount of labor. Two things are worth stressing: firstly that Dors was the only one to bring in her own hairdresser on film sets as well as on tour (Raymond, or Mr Teasy-Weasy, whom we have mentioned before) and secondly, that if you look on the Dors fan websites, there is still a lot of advice about how to replicate her hairstyle with setting lotion and "bobby pins." Clearly, her hair still has some cultural clout. And during her own time, Dors' hair became a symbol for that exotic country beyond common sense, when the time or money for extreme procedures—the dyeing, the grooming—would not be in short supply. The hair of British starlets of the period such as Sabrina, Shirley Eaton, Belinda Lee, and Liz Fraser had the same appearance, but Dors' image was also linked to a carefree consumerism and sexual voraciousness that profoundly affected its social function. Her performance style was marked by an extreme self-awareness of her own meaning: "I am, by English standards, a fairly flamboyant character—I am paid lots of money not because I look and act like the girl next door, but because my name is immediately linked with mink, fast cars and pink champagne."[4]

Clearly, Dors provided a trangressive model of femininity in the 1950s; but film culture offered alternative models too, which were not perhaps as powerful or as attractive, but that functioned as a reassuring social ballast. Actresses like June Thorburn, Susan Stephen, and Peggy Cummins were cast as essentially "nice" girls with open features and shallow-set eyes and (it can be no coincidence) similar hairstyles. They all have short, curly bobs, which are blonde and child-like. If additional evidence were needed that hairstyling was a key signifying practice in 1950s British cinema, we need only turn to the scene in *Woman in a Dressing Gown* (1957) where the heroine visits the salon. A slatternly housewife, she has lost her husband to a tidy secretary with a chignon, and she pawns her ring to afford a visit to the hairdresser. Hitherto her hair has been functional; now she is made urgently aware of its sexual value, and the hairdressing scene is one of almost unbearable humiliation. She arrives at the salon looking like herself; she leaves it looking like another. A thunderstorm tranforms her into something worse. Hair, as the building material of this scene, stimulates the viewer to run through the whole gamut of emotions—distaste, admiration, pity, despair.

By contrast, British cinema in the 1960s was largely a director's cinema, and it would be possible to categorize the use of female hairstyles by *auteur*. Since the studio system was in decline by the middle of the decade, directors began to be empowered to recruit personnel on their own behalf: designers and musicians and presumably hairstylists too. But the cinema's relation to hair fashions changed somewhat. A new fashion in the 1960s was the long, straight "hippie" style, in which the shiny texture of the hair was displayed; it signified naturalness and a degree of sexual freedom. But the dominant style in film was probably the Sassoon bob—a short cut that quoted Louise Brooks and the "flapper" look, and that was a tightly *managed* style, which required refurbishment every couple of weeks. In British cinema of the period, the "liberated" girls who created such perturbation in the texts all have these straight bobs which swing in the wind. The hair of Julie Christie in *Billy Liar* (1963), Rita Tushingham in *Smashing Time* (1967), and Vanessa Redgrave in *Morgan—a Suitable Case for Treatment* (1966) is basically *bed-hair*: tousled and unkempt, it shows what they do in private, or might have done. Of course, in other popular 1960s films such as Hammer horror films or the Bond cycle, female hairstyles are more managed, but they are still loose and "mussable."

British cinema of the 1970s underwent profound transformations, due to crises in its funding, changes in personnel, and uncertainties

about cultural change. It was a cinema that exhibited a tone of self-consciousness and irony. There is a degree of *hypersignification* in 1970s British cinema. That is to say, specific and often unpredictable areas in the film texts are given unusual prominence, and the viewers have to pick their way through a forest of signs that are heavily underscored but whose meaning is ambiguous: a single object, a fleeting expression, or a figure in a landscape. British film from the 1970s is like a palette where all the colors bleed into each other, in that the discourses from various parts of the culture get into film with greater ease than hitherto. Hence we can think about the worlds of fashion, popular music, and interior design, which have easy entry into the cinema—as the world of hairstyles does too.

In the 1970s, there were a range of popular hairstyles in Britain. One of these was the Afro, which took far longer to filter through into common practice than in America. Another was the Punk "Mohican," in which the hair over the crown is lacquered into a comb, in a style that quotes tribal practices in a postmodern manner. The dominant hairstyle in Britain in this period was the "Mullet." This hairstyle is a characteristically 1970s way of *having it all*: of being short and long, smooth and shaggy, wild and tame. And male and female too, since it is a style that is appropriated right across the board by those who perhaps ought to have known better. It is a hairstyle that requires enormous cutting skill to carry off properly. It was enthusiastically adopted by a range of film-makers. It did

not originate in the cinema, but found fertile ground there. Why was this?

It was because British cinema was, for a variety of reasons, a place where anxieties about female power could now be rehearsed. The 1970s was a period in which there had been radical changes in women's sexual lives, and the consequences of second-wave feminism were far-reaching. If we look at a range of films of the period, we can see that they manifest an *unease* about the new ways. It might be possible to argue that an ambiguity towards sexual transformation could be signaled by embracing a style that looks forward as well as back—forward to the modern, ungendered future, and back to an essential femininity. The Mullet is Janus-faced as a cultural form, and it was for that reason that it was so enthusiastically adopted by British film-makers of the period. Punk hairstyles appear rarely; they had a narrative function in *Jubilee* (1978), but that was an exceptional film.

Of course, female hair is an index of production conditions in this period as well. Consider the transformation scene in *Tommy* (1975), where the mother whips the son with her hair and throws him through the mirror, to find his own liberation. Female hair here is the hinge on which the whole narrative turns. This was only possible because of the *élan* of director Ken Russell, but such a risk-taking scene could not have been made in any other period. Or consider the female wigs in Stanley Kubrick's *Barry Lyndon* (1975), which carry very heavy narrative weight. This sublime film bore the signs of Kubrick's authorship at the level of the lighting, script, costume, and makeup. But not of the hairstyles. Kubrick was used to having total autonomy in production, but even he was defeated by the combined forces of high-status professional hairdressers, union intervention, and others. The *conception* of the wigs was the responsibility of Leonard of London, then the best-known coiffeur in town. The film union NATKE insisted on one of their members supervising the wigs' construction, since Leonard was not part of the industry. The union member had to work in a hands-on manner with Kubrick's own choice of hairdresser, Suzy Hill, as well as with Leonard, and the whole situation was complicated by some of the work having to be factored out to Stanley Hall's company Wig Creations Ltd.[5] The conflicting chains of command meant that, unusually for Kubrick, he was not in total command of this aspect of the film; the wigs escaped his control. The female wigs, particularly Marisa Berenson's, do not conform to that obsessive realism that characterizes the costumes. Instead, the hair arrangements offer a hyper-stylized, voluminous meditation on damaging imbalances between nature and culture, male and female, synthetic and natural: on how it *feels* to be part of the historical process, and to know that one's identity is a matter of contingency. That such a recognition occurs as a consequence of unforeseen circumstances during the production process is a very 1970s phenomenon.

Much remains to be done. I hope I have been able to show, through a very brief analysis of the period 1930–80, that female hair took on a variety of cultural meanings in British film, but that it was rarely as a consequence of the agency of the hairstylists themselves. I have sketched out some of the means whereby hairdressing fashions were able to filter through into the cinema, and to be modified there. A fuller picture could be created by following the work through into later periods, in order to establish whether the innovations of the 1970s were maintained, or whether they declined. But such a hair extension must be reserved for another day.

Acknowledgments

The research for some of this article was carried out under the aegis of the Arts and Humanities Research Council, which supported the University of Portsmouth project on 1970s British cinema. I would like to thank Dr Mark Glancy and Dr Justin Smith for their advice on this article.

Notes

1. Publicity material for *Madonna of the Seven Moons* held at the British Film Institute Library, London.
2. *Film Weekly*, August 23, 1933.
3. *The Morning Post*, May 10, 1937, contains a letter attacking the hairstyles in *Tudor Rose* (1936).
4. *Picturegoer*, March 24, 1956.
5. Stanley Kubrick Papers (University of the Arts, London), SK/14/2/5/3, letter

from Stanley Hall to Bernard Williams, June 13, 1973.

References

Chapman, James. 2005. *Past and Present: National Identity and the British Costume Film*. London: I. B. Tauris.

Cross, Louisa. 2008. "Fashionable Hair in the Eighteenth Century: Theatricality and Display." In Geraldine Biddle-Perry and Sarah Cheang (eds) *Hair: Styling, Culture and Fashion*, pp. 15–26. Oxford: Berg.

Douglas, Mary. 1966. *Purity and Danger: An Analysis of the Concepts of Pollution and Taboo*. London: Routledge & Kegan Paul.

Douglas, Mary. 1973. *Natural Symbols: Explorations in Cosmology*. London: Pelican Books.

Harper, Sue. 1994. *Picturing the Past: The Rise and Fall of the British Costume Film*. London: British Film Institute.

Harper, Sue. 2000. *Women in British Cinema: Mad, Bad and Dangerous to Know*. London: Continuum.

Korda, Alexander. 1937. "British Films Today and Tomorrow." In Charles Davy (ed.) *Footnotes to the Film*, pp. 171–200. London: Lovat Dickson.

Lant, Antonia. 1991. *Blackout: Reinventing Women for Wartime British Cinema*. Princeton, NJ: Princeton University Press.

Exhibition Review
Io Palmer: *Artstars*

Exhibition Review
Io Palmer: *Artstars*

The Art Gym, Marylhurst, Oregon, November 8–December 13, 2009

Braided wigs and painted strands of hair came together with a group of iconic, clothed dress forms in Io Palmer's installation, *Artstars*, at The Art Gym on the Marylhurst University campus near Portland, Oregon (Figure 1). Like an Olympic basketball dream team, five of the figures were no ordinary players. Each stood in for one of Palmer's personal art heroes, who remain unnamed. This installation had particular resonance with the gallery space itself, a former college basketball court. The gymnasium's proportions, blonde hardwood floors, and tall ceilings enhanced the artist's comparisons between art and athletic achievement.

The formal rules of sports games are usually arbitrary, yet it is necessary to perform them perfectly in order to succeed. Similarly, mastering one's skill at almost any esoteric task can lift mundane labor into art. Palmer seems to indicate that the practice of work is at least as important as art itself. Her dedication to process infuses the installation with focused intensity.

As a recurring theme, Palmer laboriously plaits hair to tell a story. The installation's converging elements pull together to create a drama with no primary narrative. The many headless hairpieces serve as markers of race, personal history, gender, and style without connoting privilege to any particular category. They are intimately specific with their hair ties and bobby pins, but still anonymous. Both the braids and the allusion to basketball refer to African American culture, although black identity politics is not the point of the installation. Instead, Palmer's vision allies with post-black art in that it extends beyond one defined viewpoint, and suggests interpretations that question the viewer's perceptions.

In a complementary role to the five *Artstars*, the six pieces in her *Janitorial Supplies* series discuss the repetitive work and dedication necessary to achieve a goal. Filling the team roster, these wig-like mop heads, their wood handles, a found metal stand and a vintage janitorial

REVIEWED BY ANNIN BARRETT
Annin Barrett teaches textiles, art, and design history at The Art Institute of Portland in Portland, Oregon.

Textile, Volume 9, Issue 3, pp. 384–389
DOI: 10.2752/175183511X13173703491225
Reprints available directly from the Publishers.
Photocopying permitted by licence only.
© 2011 Berg. Printed in the United Kingdom.

Figure 1
Artstars (installation view), 2009. Photo: Reza Safavi.

cart set up a dialogue with the *Artstars* figures. They ask, what is the connection between hard work and success? Scoring a basket in the hoop-like shape of Palmer's *Janitorial Supplies #7* (Figure 2) appears impossible, symbolizing the extreme odds against "making it" as either a professional athlete or an artist. Echoing Do Ho Suh's sculptures about individual heroes supported by a collective crowd, Palmer's installation questions how many people working behind the scenes does it take to create a "star?" What does celebrity look like? What makes the difference between fame and obscurity?

The two series, *Artstars* and *Janitorial Supplies,* are presented without hierarchy, questioning class divisions based on types of physical labor. In this post-Marxist critique, Palmer repudiates capitalism's stratified valuation of labor. We are left with pure form, the rules of the game, devoid of social status. As visual form, a crisp white service uniform in the *Janitorial Supplies* series competes well against the faintly Rococo flounced shift in *Artstars* (Figure 3). The comparison questions why is sweeping the floor any less admirable than dribbling a ball down the floor, or creating an art installation on the floor? All is work and all are equal in this assessment.

Palmer's own commitment to labor-intensive artistic process is readily apparent in these highly detailed and manipulated materials. In addition to eleven major pieces in the installation, she painted the gallery walls in

Figure 2
Janitorial Supplies #7, 2008, brace,
mop handle, synthetic hair. Photo: Reza
Safavi.

huge hanks of hair, the connecting element among all her works. Hair has long been a signifier of ethnicity, loaded with culturally imposed values of beauty and power. Lorna Simpson's *Wigs (portfolio),* 1994, investigated the intersections of race, hair, and gender by displaying photographed hairpieces as specimens of color and style. At the 2006 Whitney Biennial, Kori Newkirk's *Glint* reflected black and white cultural references from an installation built out of hair and pony beads. Similarly, the mixed black, reddish brown, and blonde braids, cornrow patterns, and wigs in *Artstars* purposefully suggest ethnic connotations but confuse any attempt at classification. In fact, all hair in this installation is synthetic, not human. Even so, our knowledge of hair is so culturally ingrained that its imitation still provokes a predictable reaction. Palmer's disembodied "fake"

Figure 3
Artstar #3, 2009, cotton, metal stand, leather, bobby pins. Photo: Reza Safavi.

hair emphasizes detachment from the head, and by extension, from any one racial identity. By objectifying hair in this manner, she disconnects some of the subconscious thought patterns about ethnic markers that are embedded in society.

Palmer also plays with the strong sense of proper placement that social conventions assign to hair. She often piles hair on the floor, sometimes being swept up by a mop. Other times, she suspends cut braids in space, or draws hair onto unusual sites. A checkerboard made on the floor out of carefully folded cloth napkins forms the base of *Artstar #1* (Figure 4). The inappropriateness of these clean linens being on the floor is accentuated by a drawing of hair clippings on each one, bringing to mind Anne Wilson's provocative household linens stitched with hair. While Wilson broke obvious taboos about hair and sexuality in public and private places,

Figure 4
Artstar #1, 2009, linen, broom, metal, drawing. Photo: Reza Safavi.

Palmer's intent is more subtle and diffuse. Curiously, all the figures in *Artstars* are gendered as women with dresses and allusions to traditional feminine domesticity. The comparison to male-dominated sports breaks down here, and it is unclear what Palmer intends by this contrast because her installation is conducive to multiple readings. Palmer's out of place objects and images symbolize a board game with no defined rules for the playing pieces in terms of gender, class or ethnicity.

Io Palmer works in a form of intense visual poetry rather than in didactic texts. The figures and objects hover dreamlike in the gallery space. As in a basketball game, the dynamic on the court constantly shifts. Hairpieces, fabric, bobby pins, and found objects are painstakingly worked into repetitive patterns. The sheer amount of labor involved commands attention and begs interpretation, but a complete understanding of the installation is confounded by the artist.

Instead, she challenges the viewer to weave their own strands of meaning from these evocative pieces, while questioning our understanding of social conditions and personal identity. Leveling the playing fields of social status, race, and gender politics, Palmer provides a setting for new arrangements of perceived value. By example, *Artstars* offers respect and incentive for the hard work involved in examining and redefining one's own relationship with society.

Exhibition Review

An Absence of Presence? *Quilts 1700–2010*

Exhibition Review
An Absence of Presence?
Quilts 1700–2010

V&A, London, March 20–July 4, 2010

The quilt world has long anticipated this exhibition and it is appropriate that visitors have come from far and wide to see this well publicized and popular exhibit. Sue Prichard and her all-female team of designers have created an eclectic mix of style and shape that elegantly showcases the V&A collection of patchwork and quilting. The V&A have researched the stories and histories behind the items to give new insight into the lives of past makers and the excellent catalog is testament to their work. The new audiovisual guides also give visitors a chance to examine the making of specific items in depth. However the sections of stitching chosen may not please everybody.

The exhibition is divided into five distinctive sections: "The domestic landscape," "Private thoughts; public debates," "Virtue and virtuosity," "Making a living," and "Meeting the past." Individual sections are enhanced by the inclusion of items from regional collections and also from overseas. Thus, we are able to make comparisons between the early pieces from the seventeenth century and those items finished this year, and to assess the changes in the craft over time. The modern items sit well with the old. However, the modern pieces chosen are recent works and it can be argued strongly that the Quilt Revival period of the last thirty years is poorly represented. We have no indication of how the work of such artists as Pauline Burbidge has changed over that time. Perhaps we are wrong to quibble over the recent past as it is but a brief period in the 300 years that the exhibition covers. It is, however, the period that as quilters we know most about. Nor does the subject matter of some of the modern pieces reflect the work of the current average or even the professional quilter. Whilst we

REVIEWED BY CAROLYN FERGUSON
Carolyn Ferguson is a freelance quilt historian, writer, and quilt maker.

Textile, Volume 9, Issue 3, pp. 390–397
DOI: 10.2752/175183511X13173703491270
Reprints available directly from the Publishers.
Photocopying permitted by licence only.
© 2011 Berg. Printed in the United Kingdom.

must admire artists like Grayson Perry and Tracy Emin for crossing over into the field of patchwork and quilting we may find the blocks of fetuses in Perry's "Right to Life" and Emin's marital bed of emotional disharmony most memorable for their subject matter. Neither are "cozy" images, but I find most disturbing and out of place, Nina Saunders's "Whispers," which depicts a distorted workbox full of babies' legs.

It is fitting that the exhibition starts both with "The domestic landscape" and with a "bed" and its patchwork hangings (Figure 1). The "bed" dates from 1730–50 and is the only set of chintz hangings from this period to be owned by a public collection. The patchwork design is one of clamshells of linen, cotton fustian, and silk, worked on a linen back. The individual motifs are made up into a design of scalloped edged diamonds that are outlined in green silk. A diary from the donor's ancestor (Baptist Noel Turner, a church of England cleric from Lincolnshire and confidante of Dr Johnson, 1739–1826) suggests that the hangings were worked on the Bedford School Estate, London, at a time when "the whole neighbourhood quilted on account of the Duke of Bedford's claim." The "Duke" refers to the Fourth Duke of Bedford, whose

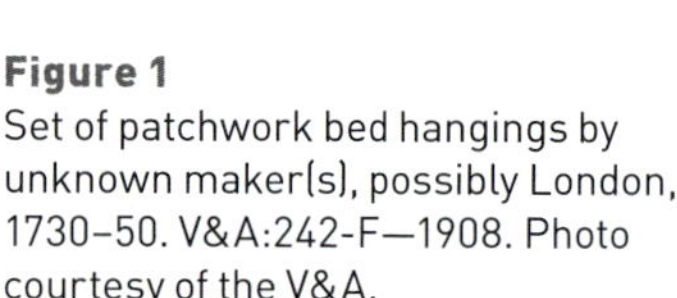

Figure 1
Set of patchwork bed hangings by unknown maker(s), possibly London, 1730–50. V&A:242-F—1908. Photo courtesy of the V&A.

ownership of large swathes of Bloomsbury and Covent Garden made the areas popular. Proceeds from the Bedford School Estate were used to fund Bedford School and to provide alms for the poor and to aid the marriage of poor women. This first section of the exhibition also shows many items from the late sixteenth and early eighteenth centuries, including a number of cot covers, bedcovers, and coverlets. The exhibition designers have cleverly allowed the visitor tantalizing glimpses of other works and one is able to view these significant historic items against the colorful backdrop of Jo Budd's two large pieces "Winter/Male" and "Summer/Female."

The second section "Private thoughts; public debates" provides a unique glimpse into the social history of the past and of the present. Its historical highlight and probably that of the whole exhibition is an early-nineteenth-century coverlet whose central panel depicts George III reviewing his volunteer troops in Hyde Park (see Figure 2). The central panel seems to have been an afterthought but is no less

Figure 2
(a) Coverlet, Patchwork of printed cottons by unknown maker, English, 1803–5. (b) Detail of stitched scene based on *Poor Jack*. V&A:T9—1962. Photo courtesy of the V&A.

(a)

(b)

significant as reviews of volunteer troops formed an important and almost daily occurrence at the time. In a country at war, these reviews importantly provided the discipline for the "territorial" army. There are forty smaller scenes shown in the border; the "eyes" show scenes of man and the military and the smaller circles show ladies at work and play. Many can be identified as being based on prints of the time. We know that needlework pictures of the 1600s were often based on engravings available in print sellers' and bookshops and that other crafts used them as inspiration too, so why not the patchworker? The scenes are shown in meticulous detail and are depicted by a combination of appliqué and embroidery. I am fascinated that the coverlet shows historical scenes spanning a range of nearly fifty years which makes the onlooker realize the importance of the print as the iconic image of the early 1800s; whether it be the death of Colonel Wolfe in Quebec (1759) or George III reviewing his troops in Hyde Park in 1799. The coverlet (Figure 2a) is shown, along with one of the border vignettes (Figure 2b), which provides an accurate reproduction of a 1790 print of a popular song title "Poor Jack" by Charles Dibdin. I was surprised that "Nelson's quilt" was not hung close by as this too provides an important reminder of the Napoleonic Wars and of this very public debate.

The "Virtue and virtuosity" section shows a number of coverlets or hangings made to celebrate the nineteenth century in pictorial form. A number are made by men and we celebrate their obvious skills in composition, design, and stitching. We also look to the religious with a Scripture Coverlet and the military with a complex geometric coverlet made in India from woolen military uniform fabric. This coverlet is attributed to Francis Brayley as maker but the very precise composition of this item, when shown alongside professional work, suggests that it was not made by an amateur and was possibly purchased by, rather than made by, Brayley. The hanging (see Figure 3) dates from 1875–85 and shows a number of appliqué pictures detailing a love story. Each picture is embroidered with a title going through the alphabet, for example Admiration, Beauty, Cupid through to Wedding and Xpression, thus providing a charming rendition of the courtship process. I feel that the modern pieces in this section coordinate better with the old with regards their color and appearance rather than their subject matter.

"Making a living" is something that all quilters aspire to do and for those of the 1930s in Wales and the North East of England it was their very bread and butter. Mavis FitzRandoph and Muriel Rose of the Rural Industries Bureau deserve much credit for keeping the craft going within the mining communities in the depression years and this section has some excellent examples of quilts made at the time. We can also compare the free-flowing quilting of the Sanderson Star of 1910–20 with Miss Nixon's red and white strippy made between 1870 and 1880. The latter shows the best quilting on a simple strippy that I have ever seen. I like the "Atelier"-type pseudo-quilt frames that show

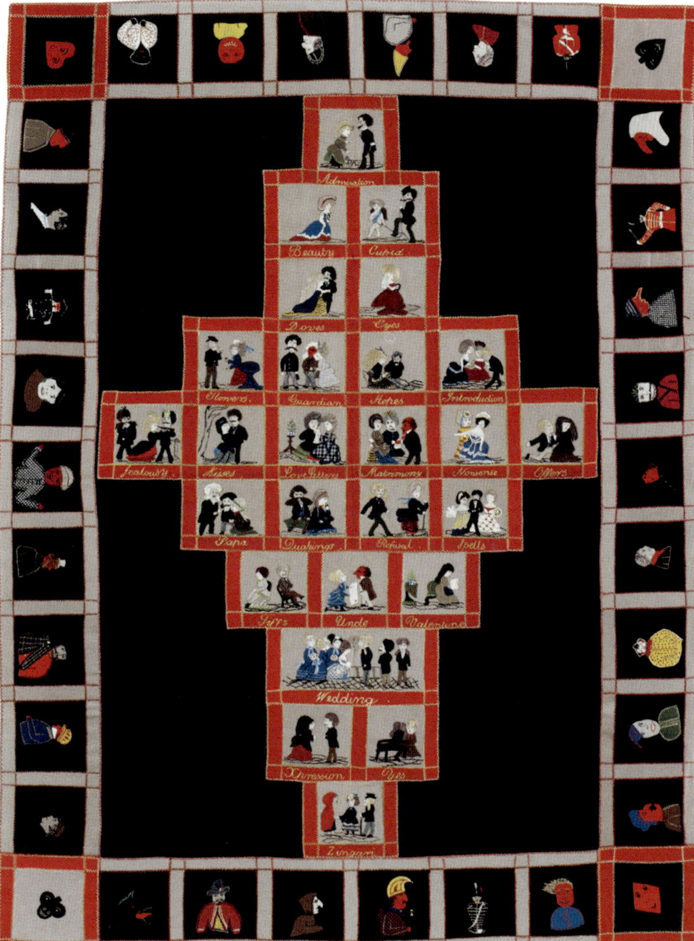

Figure 3
Cover or hanging by unknown maker,
British, 1875–85. V&A:T200—1969.
Photo courtesy of the V&A.

some of the quilts. I also like the position of the quilts on the walls, but the bare spaces make me wonder if it was not possible to show a few more items to represent the twentieth century more fully. This section emphasizes the differences in regional patterns and styles and it is good that the Irish vernacular is included. However, I am sorry that the only Scots example is made from the Timorous Beasties Glasgow Toile of 2010. We must commend the modern makers in this section,

such as Pauline Burbidge, for being able to progress their work and make a living at a time when the quilt has still to be proved a truly iconic and collectible item. Perhaps the inclusion of works by "artists" such as Perry and Emin will help to increase the status of the quilt in the world of art?

We turn the corner and the exhibition narrows to the simple white space that is "Meeting the past." Immediately we are aware of the video and the guttural sounds of the inmates of Wandsworth

Prison talking about prison and their part in the quilt that the V&A commissioned from Fine Cell Work (see Figure 4). Compare this to the very large wall-mounted Rajah quilt produced by female convicts transported to Van Diemen's land, present-day Tasmania in 1841 (see Figure 5), and the patchwork of hexagons made by Girl Guides at Changi Prison and we realize how much the physical act of sewing helps those who are incarcerated. Lady Anne Tree, now sadly deceased, the founder of

Figure 4
(a) HMP Wandsworth Quilt made by
inmates of HMP Wandsworth, London,
2010. (b) Detail of HMP Wandsworth
Quilt. V&A. Photo courtesy of the V&A.

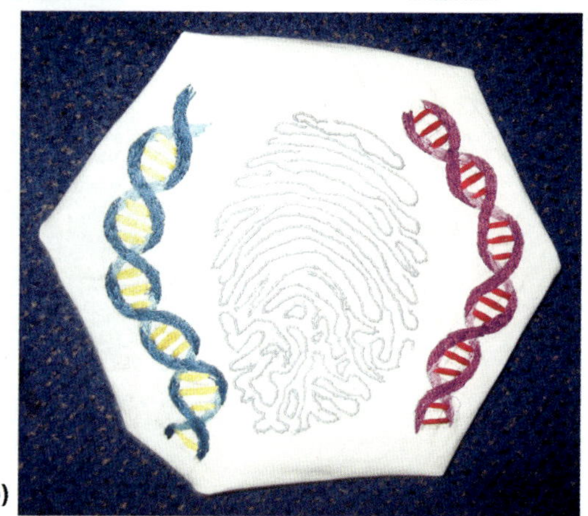

Fine Cell Work, continued the work of Elizabeth Fry, the nineteenth-century prison reformer. First Fry, then Tree realized that it was important for prisoners to learn skills to help them on both sides of the prison door. Fry was also instrumental in campaigning for better conditions on prison ships and for giving the female convicts the wherewithal to make patchworks on their voyages of transportation. It could not have been easy to sew in a moving prison and it is remarkable that such a large and complicated piece was made and survived. The three items described all provide an emotional legacy of cooperation in confinement. The exhibition ends as it began, with a bed. This time Tracy Emin creates a twenty-first-century image. As the visitor looks carefully, the rich fabrics reveal through stitch a darkness and despair. However, we do not end on a sad note. Looking back the display of quilts on "bed"-style plinths helps

Figure 5
Detail of the Rajah Quilt made by women of the convict ship *Rajah* on the journey from England to Tasmania, 1841. Collection of the National Gallery of Australia. Photo courtesy of the V&A and the National Gallery of Australia.

the viewer to visualize items within a domestic space. The windows cut from the walls give different perspectives to the items and contrast the old and the modern. Our eyes are wowed by the color of the walls and the vivid nature of the quilts gives us many memorable images. The use of audio is innovative and the Welsh quilter's lilt and the guttural prisoner's voice remind us that real people, both men and women, made these items and we celebrate them and their untold stories. I would have liked more text panels to tell the stories but we are able to read more in the expansive exhibition catalog. We thank the V&A for researching their collection and for giving the visitor a once-in-a-lifetime chance to view the items chosen to make up this significant exhibition.

Exhibition Review
I Always Live, I Never Die—An Exhibition of Contemporary Textile Art

Exhibition Review
I Always Live, I Never Die—An Exhibition of Contemporary Textile Art

Textile Museum of Oaxaca, Mexico, January 24–May 11, 2009

MEXICO: Middle-American country known for its colors ever present in the streets of towns and villages, home interiors, and artifacts . . . Perhaps this would be the way a travel guide would start out its description of this multifarious, complex country. This text attempts to wind its way through such colors.

Brought forth from deeply embedded traditions, these colors enter artistic contemporary expressions such as architecture, best exemplified in the now classic works of Luis Barragán and subsequent generations of dynamic architects; manifest in the contemporary alternative photography of Carlos Jurado, wringing color from black and white negatives created from handmade cameras, emulsions, and layers of filters; or impacting the spectator's eye in both abstract and pictorial paintings of Mexican muralists of the twentieth century, from the classic large-scale works to its contemporary women painters. It is, however, the colors of the traditional artifacts that particularly characterize such an impression of the country both for visitors and nationals—from food to the use of flowers, from paper cuttings to clay works, from celebrations to gravesites.

Additionally, it could be argued that it is the textile traditions that most strongly support such an image of a color-laden country. The enduring manifestations of a rich textile heritage and its inherent body of knowledge stem from a millenary past—where textile making was the role of practically every woman, a path assigned at birth, and where the use of colors was deemed truly significant. "Through apparently simple tools, such as backstrap loom weaving, high technical prowess was achieved" (Beauregard *et al.* 2008: 10). Textiles were considered among the highest of the artistic expressions, as impassioned

REVIEWED BY YOSI ANAYA
Yosi Anaya, PhD, is a textile researcher and artist based in Mexico.

Textile, Volume 9, Issue 3, pp. 398–405
DOI: 10.2752/175183511X13173703491315
Reprints available directly from the Publishers.
Photocopying permitted by licence only.
© 2011 Berg. Printed in the United Kingdom.

articles of trade and tribute, as intense markers of cosmic cycles and identity, stages in life, and hierarchy.

According to the introduction of the book *La Magia de los Hilos*, dedicated to the indigenous textiles traditions currently present in the State of Veracruz:

> *With the Conquest, many forms of textiles disappeared, such as the case of cloths used in temples, markets, celebrations, clothing and religious vestments both for priests and royalty, banners, warrior suits, codexes and for effigies of the deities among other things.* (Beauregard *et al.* 2008)

Needless to say, along with the repression of a wide variety of cultural expressions, the annihilation of peoples from the sixteenth century onward served its purpose. The textiles that pervaded were precisely those attached to daily living and wearing. That is, all the textiles associated with ritual use of cloth applied within the former beliefs and previous social structures were systematically eliminated. And though limited mostly to the indigenous population and despite meeting continual historical disdain, the continual use of certain textiles, attached to daily living and wearing, has endured to the present. Throughout the colonial period, a European view of cloth pervaded—that it could be woven in great widths and lengths in order to be cut. Its manufacture and use broadened indigenous skills to include male craftsmanship in textile making with large treadle looms, so intensifying production,

which would be selectively assimilated and continually reinterpreted by the native peoples.

The region of Oaxaca, a southern state of Mexico, is characterized by such a variety of traditions. And it was indeed Oaxaca that produced many of the precious colors coveted by Europe—particularly indigo and cochineal. Oaxaca is still considered to be the hallmark region of Mexico's historic colorants, kept alive in the indigenous cultures,[1] and to be home to the richest textile traditions. Formerly, these native coloring agents were applied not only to textiles but also to buildings, murals, homes, temples, ceramics, feather and woodwork, sculpture, offerings, food, books, etc.

The Museum

In the first decade of the twenty-first century, a much-needed museum developed from an original project in the capital city of the state of Oaxaca. Conceived by artists, patrons, and collectors, it opened in May 2008. The Museo Textil de Oaxaca (MTO) set about the task of maintaining active communication between traditional knowledge of textiles and its living exponents linked to the promotion, conservation, study, and sharing of the textiles in its collections. To further the revitalization of Oaxaca's textile culture, the MTO's program involves openness— rendering homage to salient weavers, dyers, traditional cultures, and projects through a lively series of events and well-curated exhibitions. The MTO has become a second home to many of Oaxaca's textile artists. Its president, Dr María Isabel Grañen Porrúa, has made this manifest: "Open the doors, you will be welcomed to this cocoon that shows the beauty born of threads—this is your home and a gift to Mexico" (Porrúa 2008: 36).

Although its collected works (now numbering 4,761) mostly focus on Oaxaca's indigenous groups as well as other Mexican native peoples, textiles from different world cultures have been added to augment the collections. It may seem that the focus lies specifically on upkeep and furthering skilled traditions; however, the MTO embraces in its concept of textile art *all artistic textile manifestations*—from historic to deeply traditional textiles to contemporary textile expressions. Its first exhibition of contemporary textile art took place from January 24 to May 11, 2009. This pioneering show was curated by textile artist Androna Linartas, organizer of other notable textile exhibitions in Mexico City in previous decades.

Yo siempre vivo, yo nunca muero (*I Always Live, I Never Die*) at the MTO marks the first national group exhibition in the country after a number of years of silence—nearly a decade—exclusively showcasing contemporary textile art, with the exception of Mexican artists exhibiting abroad in solo shows, biennials, and triennials. For the young museum, this was an important step, given the special criteria of its internal curatorship program, which focus on a theme, motif, dye, technique, or particular weaver in Oaxaca and world textiles.

On this occasion, the museum invited Linartas as external curator to form a free-format textile exhibition that would combine the varied works of eight artists living in Oaxaca with fifteen carefully selected artists from the rest of Mexico. However, it was not a completely limitless participation, for each work had to pertain to the central theme of the show: *Oaxaca*—be it about its landscape, flavors, rituals, dyes, textiles, traditions, festivities, colors, and so forth—a land where a sense of belonging and identity is underscored.

Set in the great T-shaped hall, Linartas worked with the museum's team of museographers and Director Ana Paula Fuentes to coalesce the varying discourses of each artist into one cohesive and successful exhibition. The challenge for Linartas was to make a respected and significant show to counter the decade-old gap without a major contemporary textiles exhibition in the country, despite Mexican artists participating individually and internationally. Mexico, bound to the economic power-play of the world, is a country with many feet (metaphorically, rather than applying the generally accepted term multifaceted). It has a foot in modernity, another in tradition, another two in postmodernity, and several others in deep, unsolved problems. Its art scene, though provocative and innovative in its proposals, has, as in so many other countries, made it obvious that textiles have a *different* space among the arts—hardly a space. This exhibition thus presents a challenge both for the "Western" art panorama in Mexico as well as for the museum, in staging a show exclusively of contemporary textile art while maintaining its open communication with the people of Oaxaca, who visited, highly interested.

Walls, ceiling, floor, and tables were occupied—by a variety techniques. The exhibition was a dialogue between experimentation and tradition in individual proposals or personal discourses. Although there were numerous works that were effective and evocative, a few, as in any group show, were less interesting, styles already very much seen, as the challenge was left open to each selected artist to work specifically on the exhibition theme. A thin thread links some of the works, while others contrast with each other, despite the very broad underlying theme. It can be discerned that as much as there are artists who work their proposals mainly through a textile technique—shaping and crafting their language through their skill—there are those who are beyond a textile technique, where lies the concept of the discourse and personal search. Thus, *Yo siempre vivo, yo nunca muero* brought together a multiplicity of discourses expressed through textiles and it could be assessed that statements on the notion of *Oaxaca* are indeed multiple. The exhibition design gave an equally important space to each of the artists, making not only an interesting, harmonious whole (although tightly packed) but also allowing for individual impact that communicated both to Western and indigenous visitors.

The Works

In the exhibition, three of the four male artists employ tapestry. Two renowned Oaxacan artists have delved into working with traditional weavers, to produce tapestry works in the classic form, based on their paintings turned into *cartons*.

Here the ability of their Oaxaca weavers to interpret the quality of their originals through the use of weaves and specially colored yarns is evident. Known for his patronage of the arts in Oaxaca, and one of the founders of the MTO, Francisco Toledo has long worked with Zapotec weavers to produce his Gobelin-like tapestries where large floor looms are used. In his *El murciélago* (*The Bat*), the central figure has their head hanging down as when sleeping but nonetheless with wings open. Rodolfo Morales's work *Adivinanza de la Esperanza* (*Riddle of Hope*) is a large tapestry of one of his complex paintings, also woven by a Zapotec master weaver, who interpreted the tones of his elaborate original through a rich use of textures and mixed colored yarns. On the opposite bench, however, is Gabriel Canales, a tapestry artist himself, who prefers to weave in small formats to construct a large work through hand-size units, delving into the painters' realm presenting three wide woven paintbrushes; this is part of his investigation into the quotidian aspects of color and textiles in the "blue-collar trade" of the "broad-brush" painter, a counterpoint to the fine artist.

Two works focus on Zapotec traditional symbols and both do so in contrasting black and white. Trine Ellitsgaard, a weaver from Denmark settled in Oaxaca, works with fibers native to Mexico. In Ellitsgaard's untitled sisal weaving, the repeating star design of the Isthmus Zapotec *huipiles*,[2] usually stitched continuously in sequences to form bright contrasting bands, is blown up to large proportions. Set into a minimalist aesthetic, the design is translated from stitching to weaving and isolated from its usual cultural context. Because the appropriation of indigenous cultural patrimony is disputable, the excellently executed star design here threads on uneasy, uncomfortable acceptance. Teresa Olabuenaga also centers on a Zapotec motif from the pre-Columbian past: the stepped fret in *El Muertito Enfiestado* (*The Feasted Dead*), taken from the carved stone friezes at Mitla, the extraordinary pre-Hispanic Zapotec ceremonial city. But rather than placing the motif in continuous sequences as on Mitla's walls, Olabuenaga uses it repeatedly, dangling almost at random, in small individual paper prints. The work's wooden boxlike frame limits the freedom of the cords and fret motif within a dark space. The importance of this symbol in diverse textile languages harks back to the theme of the MTO's first exhibition, *From Mitla to Sumatra, the Art of the Woven Fret*, curated by Alejandro de Ávila.[3]

Sara Corenstein takes her own personal texts—journals dating from more than fifteen years ago—cuts the pages into strips and weaves them. These texted weaves in subtly varying colored inks recall her past in a Benjaminian meticulousness of recycled continuity. Ornella Ridone also incorporates a lived object—her bed sheet. With tears, snags, repairs, burns, stitching, embroidery, and layers of painted patches, it is a poetic evocation of recurring activities rather than recurring sleep. Rosa Luz Marroquín too recycles and gives a new light to her long-time worn traditional textiles, now cut up into pieces which she dyes in deep reds and purples, sews them up to compose impressive large ceremonial robes, adorned with beads, stitching, silver religious ex-votos, and occasional feathers. Demian Flores on the other hand, in *Estandartes* (*Banners*), intervenes in the discarded festive objects of popular culture: religious procession banners form the backdrop for his intervention with shifted, humorous iconography.

Georgina Toussaint is drawn to the marketplace where one can purchase daily-use implements, such as plastic ropes to set up clotheslines. She weaves these colored ropes into a shape that insinuates the hidden primordial woman's body caught in the tediousness of daily chores. Both Mahia Biblos and Rowena Galavitz have worked on tank-top T-shirts. Rowena gold-leafed cornhusks and carefully stitched into the tank-top T-shirt; while Biblos punctured the T-shirt with sharp pencils that also had bold texts written on the garment's surface. The usual urban second skin is painfully pierced and violently stitched and marked. In contrast with Los Danzantes (*The Dancers*) by Sara María Terrazas, evoking four striated bodies in delicate paper and threadwork.

Although Marcela Gutiérrez names her work *Caminos de Oaxaca* (*The Roads of Oaxaca*), rather than being earthbound it is aerial—hanging high like a kite complex or colored clouds in an Oaxaca sunset. Maya Amrein, however, alludes to death as a barren ground in her floor piece of dark, almost sinister, serpent-like vegetable fibers, contrasting with Yosi Anaya's installation, *Ríos de mis Ancestros* (*Rivers of my Ancestors*), that pays tribute to the spirits of

her predecessors in their afterlife in traditional bark paper, cloth strips, and native historic dyes (indigo, *cochineal, cempasuchil, and palo brazil*) sewn together to give the effect of their travel through the waters of a river, pertaining to Mexico's traditional annual practice of tribute to the dead.

Mira Jokic's pebbles, a printed and felted cloth sculpture, are an enlarged detail of something that could be from a riverbed in any part of the planet. Maddalena Forcella achieves the true carmine hue of cochineal in felted wool, almost reminiscent of Scythian felt and cave paintings. Laurie Litowitz presents an almost sterile and enigmatic long table encased in glass and containing glass flowers, and crocheted silver wire. This ambiance of metallic shine and transparent whiteness contrasts with Enna Negrón's floor work of similar proportions, presenting colorful woven and embroidered flowers of traditional blouses floating on a long tense warp. *A Donaji* (*To Donaji*), however, lies beyond the apparently sweet presentation of romance that flowers may evoke. Though these characteristically embellish a woman's indigenous dress, the stray flowers are now laid out at ground level, representing many trodden indigenous women that nonetheless radiate life and color.

The characteristic dress of Oaxaca's Isthmus women is reinterpreted in metallic mesh against the backdrop of a white wall in María Luisa Simón's *Ellas Caminan en Verso* (*They Walk in Verse*). A woman's body within is insinuated by the creases and folds in the colors of the different

metals—copper and bronze—while subtle embroidery hues soften the metallic harshness of the material. This work, above all the others, captures the exhibition's theme—in dreaming of Oaxaca. The female spirit in the inhabited dress of the ever-present, dark-skinned, sensuous Isthmus woman who dominates marketplaces floats as in a dream pervading: *Whilst you sing to me, I will always live, I never die . . .* Hence, the exhibition's title, taken from a Zapotec poem and song—*La Martiniana* by Andres Henestrosa—resonantes within the exhibition. Linartas (2009: 16) writes, "Textile Art has not died, but on the contrary, it has always been alive, in constant evolution."

How to interpret *Oaxaca*? The artists faced the challenge of whether to dwell on its colors or to use its iconography, to take from its traditions or from its convergence of materials, techniques, abstractions, images. The question arises as to how much an artist can appropriate from traditional cultures in resistance. Whether the use of the colors, the designs, the cultural manifestations, even the spirit of cultural confluences can be used advantageously by Western artists in the common act of random appropriation—without permission, overstepping collective cultural rights.

This is a time of apertures, of undoing limits, a time when traditional cultures are opening towards promotion of their art products and practices. But at the same time, native peoples demand respect for their expressions, art, and organizations for which they have struggled long to maintain.

At times an appropriation of an element from another culture may serve in its promotion. But at others, it could de-contextualize, make its significance seem banal, overexploiting a practice or a valuable symbol. However, this exhibition does not even come close to bordering on any exploitation of indigenous peoples. On the contrary, it exalts them. It captures some of the multiple flavors, colors, and nuances of Oaxaca's confluence of evolving cultural practices, to complement them.

Additional events related to the exhibition included a papermaking workshop by participating artist Maya Amrein, a weaving course given by the curator, and an artist's talk by this review's author. Linartas's intention—"that this event be the beginning of a new and very important chapter in the history of Textile Art in Mexico" (Linartas 2009: 16)—will certainly resound in coming years, as the MTO has now opened a new permanent space, *Cubo Textil Contemporáneo*, for continuous contemporary displays. Important is the fact that the impact of *Yo siempre vivo, yo nunca muero* has transcended locality. The show reopened in a different format at the renowned Museo Franz Mayer in Mexico City in 2010 and will most probably go on to other museums throughout Mexico. The presence of Oaxaca thus has doubly resounded in the nation's capital, re-nourishing the MTO's pioneering efforts.

Notes

1. Major indigenous groups in Oaxaca include Mixtec, Zapotec, Huave, Mazatec, Chinantec, Chatino, Nahua, Mixe, Trique, Tacuate, Amuzgo.

2. Meso-American women's upper garment of varying lengths composed of straight woven rectangular panels. The Isthmus *huipiles* are more like square-cut, snug blouses than the draping tunics in other regions.

3. de Ávila researched the recurring multifaceted symbol across textile cultures. His earlier treatise linked the stone designs of Mitla to widespread preceding weaving traditions of this Mesoamerican region (see de Ávila 1998).

References

Beauregard, L., Aquino, L. and Anaya, Y. 2008. *La Magia De Los Hilos: Arte y tradición en el textil de Veracruz*. Xalapa: Editora Gobierno del Estado.

de Ávila, Alejandro. 1997. "Un Huipil Colorado: Tiempos del textil oaxaqueño." In *Historia del Arte de Oaxaca, Vol. III—Arte Contemporáneo*, pp. 217–62. Oaxaca: Gobierno del Estado de Oaxaca-Instituto Oaxaqueño de las Culturas.

de Ávila, Alejandro. 1998. *El Arte de Oaxaca, Vol. II*. Oaxaca: Gobierno del Estado de Oaxaca.

Galimberti, Alexandra (coord.) *et al.* 2008. *Oaxaca en Femenino: 40 mujeres en las artes visuales*. Mexico City: CONACULTA-Consejo Nacional para la Cultura y las Artes.

Grañen Porrúa, María Isabel. 2008. *El museo textil de Oaxaca, un regalo para México*. Oaxaca: ADABI-Apoyo al Desarrollo de Archivos y Bibliotecas de México.

Linartas, Androna. 2009. *Yo siempre vivo, yo nunca muero* (exhibition catalog). Oaxaca: Museo Textil de Oaxaca.

Notes for Contributors

- Articles should be approximately 25 pages in length and *must* include a three-sentence biography of the author(s) and a short abstract (150–200 words).
- Interviews should not exceed 15 pages and do not require an author biography.
- Film, exhibition and book reviews are normally 500 to 1,000 words in length.
- The Publishers will require a disk as well as a hard copy of any contributions (please mark clearly on the disk what word-processing program has been used).

From time to time, *Textile: The Journal of Cloth and Culture* plans to produce special issues devoted to a single topic with a guest editor. Persons wishing to organize a topical issue are invited to submit a proposal which contains a hundred-word description of the topic together with a list of potential contributors and paper subjects. Proposals are accepted only after a review by the journal editors and in-house editorial staff at Berg Publishers.

Manuscripts

Manuscripts should be submitted to Amanda Stonham, Editorial Assistant (c/o catherine.harper@uel.ac.uk). Manuscripts will be acknowledged and entered into the review process discussed below. Manuscripts without illustrations will not be returned unless the author provides a self-addressed stamped envelope. Submission of a manuscript to the journal will be taken to imply that it is not being considered elsewhere for publication, and that if accepted for publication, it will not be published elsewhere, in the same form, in any language, without the consent of the editor and publisher. It is a condition of acceptance by the editor of a manuscript for publication that the publishers automatically acquire the copyright of the published article throughout the world. *Textile: The Journal of Cloth and Culture* does not pay authors for their manuscripts nor does it provide retyping, drawing, or mounting of illustrations.

Style

US spelling and mechanicals are to be used. Authors are advised to consult *The Chicago Manual of Style (16th Edition)* as a guideline for style. *Webster's Dictionary* is our arbiter of spelling. We encourage the use of major subheadings and, where appropriate, second-level subheadings. Manuscripts submitted for consideration as an article must contain: a title page with the full title of the article, the author(s) name and address, a short abstract (150–200 words) and a three-sentence biography for each author. Do not place the author's name on any other page of the manuscript.

Manuscript Preparation

Manuscripts must be typed double-spaced (including quotations, notes, and references cited), one side only, with at least one-inch margins on standard paper using a typeface no smaller than 12pts. The original manuscript and a copy of the text on disk *(please ensure it is clearly marked with the word-processing program that has been used)* must be submitted, along with black and white original photographs (to be returned). Authors should retain a copy for their records. Any necessary artwork must be submitted with the manuscript.

Footnotes

Footnotes appear as "Notes" at the end of articles. Authors are advised to include footnote material in the text whenever possible. Notes are to be numbered consecutively throughout the paper and are to be typed double-spaced at the end of the text. (Please do not use any footnoting or end-noting programs which your software may offer as this text becomes irretrievably lost at the typesetting stage.)

References

The list of references should be limited to, and inclusive of, those publications actually cited in the text. References are to be cited in the body of the text in parentheses with author's last name, the year of original publication, and page number – e.g., (Rouch 1958: 45). Titles and publication information appear as "References" at the end of the article and should be listed alphabetically by author and chronologically for each author. Names of journals and publications should appear in full. Film and video information appear as "Filmography." References cited should be typed double-spaced on a separate page. References not presented in the style required will be returned to the author for revision.

Tables

All tabular material should be part of a separately numbered series of "Tables." Each table must be typed on a separate sheet and identified by a short descriptive title. Footnotes for tables appear at the bottom of the table. Marginal notations on manuscripts should indicate approximately where tables are to appear.

Figures

All illustrative material (drawings, maps, diagrams, and photographs) should be designated "Figures." They must be submitted in a form suitable for publication without redrawing. The publishers encourage artwork to be submitted as high-resolution (600 dpi or above) electronic files (TIFF or maximum quality JPG format) on disk or via email. Hard-copy drawings should be carefully done with black ink on either hard, white, smooth-surfaced board or good quality tracing paper. Computer-generated drawings of publishable quality are also welcomed. Color photographs are encouraged by the publishers. Whenever possible, photographs should be 8 × 10 inches. All figures should be numbered consecutively. All captions should be typed double-spaced on a separate page. Marginal notations on manuscripts should indicate approximately where figures are to appear. While the editors and publishers will take every care in protecting all figures submitted, they cannot assume responsibility for their loss or damage. Authors are discouraged from submitting rare or non-replaceable materials. It is the author's responsibility to secure written copyright clearance on *all* photographs and drawings that are not in the public domain. Copyright should be obtained for worldwide print and electronic publication rights.

Criteria for Evaluation

Textile: The Journal of Cloth and Culture is a refereed journal. Manuscripts will be accepted only after review by both the editors and anonymous reviewers deemed competent to make professional judgments concerning the quality of the manuscript. Upon request, authors will receive reviewers' evaluations.

Copies

On publication, authors will be sent a PDF of the final version of their article for personal use only. Authors are also entitled to a complimentary copy of the issue they contributed to. Details on how to obtain it will be sent upon publication. Additional copies of the issue can be purchased at a discounted rate from Berg.